NCRP REPORT No. 115

RISK ESTIMATES FOR RADIATION PROTECTION

Recommendations of the
**NATIONAL COUNCIL ON RADIATION
PROTECTION AND MEASUREMENTS**

Issued December 31, 1993

National Council on Radiation Protection and Measurements
7910 Woodmont Avenue / Bethesda, Maryland 20814-3095

LEGAL NOTICE

This Report was prepared by the National Council on Radiation Protection and Measurements (NCRP). The Council strives to provide accurate, complete and useful information in its reports. However, neither the NCRP, the members of NCRP, other persons contributing to or assisting in the preparation of this Report, nor any person acting on the behalf of any of these parties: (a) makes any warranty or representation, express or implied, with respect to the accuracy, completeness or usefulness of the information contained in this Report, or that the use of any information, method or process disclosed in this Report may not infringe on privately owned rights; or (b) assumes any liability with respect to the use of, or for damages resulting from the use of any information, method or process disclosed in this Report, *under the Civil Rights Act of 1964, Section 701 et seq. as amended 42 U.S.C. Section 2000e et seq. (Title VII) or any other statutory or common law theory governing liability.*

Library of Congress Cataloging-in-Publication Data

Risk estimates for radiation protection / recommendations of the National Council on Radiation Protection and Measurements.
    p. cm.—(NCRP report; no 115.)
  "Issued December 31, 1993".
  Includes bibliograpical references and index.
  ISBN 0-929600-34-7
    1. Tumors, Radiation-induced—Epidemiology. 2. Ionizing radiation—Safety measures. 3. Health risk assessment. I. National Council on Radiation Protection and Measurements. II. Series.
RC268.55.R57 1993
616.9'89705–dc20   93-44242
                                                                        CIP

Preface

The United Nations Scientific Committee on the Effects of Atomic Radiation (UNSCEAR, 1988) reviewed the radiation epidemiology data of the Japanese atomic-bomb survivors for the period of 1950 through 1985. Additionally, the National Academy of Sciences/ National Research Council (NAS/NRC, 1990) has reviewed the radiation epidemiology data of the Japanese survivors for the same time period. Each of these committees made separate independent evaluations of the data and each provided radiation risk estimates for the general population and a worker population. However, neither report made specific recommendations on the method of converting their risk estimates for high-dose and high-dose-rate exposure to low doses and dose rates applicable to radiation protection. Additionally, neither report addresses the translation of the risks from a Japanese population to a United States population.

The National Council on Radiation Protection and Measurements (NCRP), in preparation for making new radiation protection recommendations, commissioned the work of Scientific Committee 1-2, *The Assessment of Risk for Radiation Protection Purposes* to review the work of UNSCEAR (1988) and NAS/NRC (1990). This Report provides the results of this review and is the bases on which the recommendations in NCRP Report No. 116, *Limitation of Exposure to Ionizing Radiation,* are made. Serving on Scientific Committee 1-2 were:

R. J. Michael Fry, *Chairman*
Oak Ridge National Laboratory
Oak Ridge, Tennessee

Members

Seymour Abrahamson
Radiation Effects
  Research Foundation
Hiroshima, Japan

William J. Schull
Radiation Effects
  Research Foundation
Hiroshima, Japan

Jacob I. Fabrikant*
University of California
Berkeley, California

Warren K. Sinclair
National Council on
  Radiation Protection
  and Measurements
Bethesda, Maryland

Charles E. Land
National Cancer Institute
Bethesda, Maryland

Arthur C. Upton
New York University
  Medical Center
Tuxedo, New York

Roger O. McClellan
Chemical Industry Institute
  of Toxicology
Research Triangle Park,
North Carolina

Edward W. Webster
Massachusetts General
  Hospital
Boston, Massachusetts

NCRP Secretariat

William M. Beckner, *Scientific Staff Assistant*
Cindy L. O'Brien, *Editorial Assistant*

The Council wants to thank the Committee for its work.

Charles B. Meinhold
President, NCRP

Bethesda, Maryland
October 26, 1993

*deceased

Contents

1. Summary

This Report provides a critical examination of the information about risks from exposure to ionizing radiation that was considered necessary before an update of the National Council on Radiation Protection and Measurements' (NCRP) Report No. 91, *Recommendation on Limits for Exposure to Ionizing Radiation*, could be made. The Report is concerned with the stochastic effects of radiation, namely cancer and genetic effects and it also discusses the effects of radiation on the developing brain, which are considered to be deterministic effects. The main task was an assessment of reports prepared by the United Nations Scientific Committee on the Effects of Atomic Radiation (UNSCEAR, 1988) and the Committee on the Biological Effects of Ionizing Radiations (BEIR V) (NAS/NRC, 1990) as well as the identification of findings that should be considered in the preparation of the report *Limitation of Exposure to Ionizing Radiation*, NCRP Report No. 116 (NCRP, 1993). The International Commission on Radiological Protection's (ICRP) Publication 60 (ICRP, 1991) is not specifically reviewed; however, material prepared for that report by members of the committee responsible for this Report have been incorporated.

The data for cancer mortality in the atomic-bomb survivors are the major source on which the risks of radiation-induced cancer are based. In 1977, ICRP based its recommendations for radiation dose limitation on the risk estimates of UNSCEAR (1977). The estimate for radiation-induced excess lifetime cancer mortality was based on the data for induction of leukemias and an assumption that the ratio of solid cancer to leukemias was about five. UNSCEAR (1988) reported risk estimates for lifetime excess cancer mortality and life shortening, the excess risk for specific individual organ sites and other cancers (the so-called remainder group) for the years 1950 to 1985 using the new dosimetry, 1986 Dosimetry System (DS86). The selection of the cancer sites was restricted to those sites for which a significant increase in mortality with dose could be demonstrated.

The BEIR V Report (NAS/NRC, 1990) considered that the numbers of excess cases for many sites were not sufficient to allow the necessary stratification into different dose, age and time categories. Therefore, they restricted their estimates of risk to leukemias, breast cancer, cancers of the respiratory and digestive systems and thyroid

1

cancer. The risks for these sites and systems were modeled individually. The estimate of the total excess lifetime cancer mortality took into account cancers at other sites.

Despite investigations of a number of irradiated populations such as the patients with ankylosing spondylitis (Darby *et al.*, 1987), those with cervical cancer treated with radiotherapy (Boice *et al.*, 1987) and the radiation worker studies that date back to the 1970s (Gilbert and Marks, 1979) and continue to this day in the United States, Great Britain and elsewhere, the data for the atomic-bomb survivors remain the main basis for ionizing radiation risk estimates. However, the BEIR V Report (NAS/NRC, 1990) did use data from six other studies in their estimates of risks for specific sites.

UNSCEAR (1988) estimated the risk of cancer for organ doses of 1 Gy using a neutron relative biological effectiveness (RBE) of one. A linear-dose response for solid cancers and a linear-quadratic model for leukemia were used to analyze the data of the "not in city" and the 0.01 to 4 Gy dose groups. In the estimates of risk for the general and worker populations by UNSCEAR (1988) and BEIR V (NAS/NRC, 1990), no allowance was made for low-dose rates or low doses.

The BEIR V Report (NAS/NRC, 1990) also used a linear-quadratic model to fit age-specific mortality data for leukemias and a linear-dose response for solid cancers in their estimates of risk of excess cancer and life shortening in persons exposed to a single acute exposure of 0.1 Sv or a continuous lifetime exposure to 1 mSv y^{-1} and 0.01 Sv y^{-1} from age 18 to 65 y of age. The estimates were based on data for exposures at a high-dose rate with no reduction of the risk of solid cancers for exposures at low-dose rates and using a neutron RBE of twenty.

In 1985 over 60 percent of the atomic-bomb survivors were alive, the majority of whom had been in the younger age groups at the time of the bombings (ATB). Thus, it has been necessary to project the risk estimates over time to obtain estimates of excess lifetime cancer mortality. UNSCEAR (1988) and BEIR V (NAS/NRC, 1990) differed in their selection of a risk projection model. UNSCEAR (1988) chose a period of 40 y after exposure as the duration of excess mortality from leukemias and lifetime for solid cancers. Risk estimates were made on the basis of constant additive and constant relative or multiplicative risk models. The latter was preferred and it gave a lifetime risk estimate of excess cancer mortality based on age-averaged risk coefficients for both sexes of 7.1 × 10^{-2} Gy^{-1}, compared to 4.5 × 10^{-2} Gy^{-1} for the additive model. The approach that UNSCEAR (1988) used assumes that the minimum latent period is independent of age at exposure and dose and can be grouped into two values, 2 y for all types of leukemias and bone cancer, and 10 y for all solid cancers. Similarly, the duration of excess risk is assumed to be the same for all solid cancers and independent of the age at exposure. The BEIR V Report (NAS/NRC, 1990), noting the apparent decrease in the risk of cancer in the cases of ankylosing spondylitis after about 25 y reported by Darby et al. (1987), considered it necessary to model how the risk varied with time and with age at exposure.

The approach to life-table calculations by the two committees were different. UNSCEAR (1988) calculated the number of cancer deaths attributable to radiation in a standard population and this is different from the number of excess deaths in a population used by BEIR V (NAS/NRC, 1990). The BEIR V estimate excludes those who die of a radiation-induced cancer who would have otherwise died at a later time from a cancer of natural occurrence.

Although the two committees used different methods of analyses, the estimates of lifetime cancer risks for acute whole-body exposure to low-LET radiation were comparable, 7.1×10^{-2} Sv^{-1} (UNSCEAR, 1988) and 8.8×10^{-2} Sv^{-1} (NAS/NRC, 1990). The difference between these two estimates is smaller than the difference between the UNSCEAR (1988) estimates of lifetime risk of fatal cancer derived with age-averaged risk coefficient of 7.1×10^{-2} Sv^{-1} and with age-specific risk coefficients of 11.0×10^{-2} Sv^{-1}.

Neither committee applied a dose-rate reduction factor for low-dose rates because of the lack of data for humans, but both committees indicated a factor of between two and ten might be appropriate. UNSCEAR (1988) assumed an RBE for neutrons of one but BEIR V (NAS/NRC, 1990) chose a value of twenty.

The conclusion of the two committees is that the new estimates are significantly greater than those of BEIR III (NAS/NRC, 1980) and UNSCEAR (1977), even when a dose-rate-effectiveness factor of two to three is applied. The risk estimates are based on a total of about 5,900 cancer deaths, about 344 of which are attributed to radiation. The number of excess cancer deaths is small when the stratification of the data for sex, age at exposure, city and dose group is carried out.

The estimates of genetic risk have not changed greatly over the last decade. The doubling dose (DD) for genetic diseases that cause morbidity or mortality in humans is now estimated to be about 1.7 to 2.2 Sv or about 3.4 to 4.4 Sv for exposures at low-dose rates. These values are three to four times greater than the DD obtained from studies of the mouse. Because of the large uncertainties in both estimates and the desire to provide adequate protection, a risk of severe hereditary effects for the general population (for all generations) of 1×10^{-2} Sv^{-1} should provide a reasonable basis for dose limitation. There remains an important uncertainty in the estimate of radiation-induced genetic effects. This uncertainty pertains to the estimate for the so-called multifactorial diseases. These diseases include cancer and heart disease. The mutational component of these diseases is not known with any accuracy, nor is the effect of radiation on this mutational component known. For these reasons, neither UNSCEAR (1988) nor BEIR V (NAS/NRC, 1990) derived a risk estimate for this category of diseases.

The severity of radiation effects on the brain is dependent on gestational age at exposure. Four time periods based on the stages of development and measured from time of fertilization, have been used for estimating the sensitivity for damage to the brain in relation to gestational age. The second period, 8 to 15 weeks, is the period of greatest risk of severe mental retardation. This is the period of rapid proliferation and of migration of neurones. There is a smaller risk as a result of exposure between 16 to 25 weeks gestational age and no apparent risk of severe mental retardation if the exposure is after the 25th week. In the first period, 0 to 7 weeks, there may not be survival to an age that mental retardation could be recognized. The data for the 8 to 15 week group, based on a linear-dose response, yield a probability of induction of severe mental retardation of 0.4 per Gy. The question is whether a threshold exists. A threshold is the earmark of deterministic effects. In the case of exposure in the 16 to 25 week period of gestation, a threshold of 0.23 Gy is estimated at the lower 95 percent confidence limit. In the case of the 8 to 16 week period, the data can be fitted by quadratic and linear responses and the possibility of a small but significant threshold ranging from 0.12 to 0.23 Gy exists. However, for radiation protection purposes, it is considered prudent to use a risk probability of 0.4 per Gy. Less severe impairment of mental ability has also been found in children exposed *in utero*. The effects are reflected in a dose-related decrease in intelligence test scores, with an estimated shift of 30 units per Sv in intelligence quotient (IQ), impaired scholastic performance, development and even seizures.

Radiation protection is concerned with the total potential detriment from exposure. Based on the reports of UNSCEAR (1988), BEIR V (NAS/NRC, 1990), ICRP (1991) and other studies, the NCRP, in this Report, has estimated the total detriment resulting from exposure to low-LET radiation for the general population to be about 7×10^{-2} Sv^{-1} and for the working population about 6×10^{-2} Sv^{-1}. The total detriment includes fatal and nonfatal cancer, severe hereditary effects and life shortening which have been estimated separately. The major contribution to the total detriment is from fatal cancer of about 4.0 and 5.0×10^{-2} Sv^{-1} for the worker and general population, respectively.

Risk estimates of radiation effects are being made with increasing precision. However, it is important to consider the many sources of uncertainty and the probability of the reduction in the level of uncertainty. Currently, the estimation of radiation risks for both the general and worker population depends largely on the data for the atomic-bomb survivors. There are concerns about the appropriateness of a population with a distinct distribution of naturally occurring cancers exposed in war time at very high-dose rates of radiations for the assessment of risk to populations in the western world and exposed either to small fractions or protracted irradiation at very low-dose rates.

The sources of uncertainty include:

1. Dosimetry, which has involved complex reconstruction of the characteristics of the exposure and the location of the exposed persons, presents both systematic and random errors.
2. The quality of the data base, including accuracy of diagnosis.
3. The choice of a risk-projection model. This may represent a significant uncertainty since over 60 percent of the atomic-bomb survivors are still alive.
4. Extrapolation of the risk estimate based on data for high doses incurred at a high-dose rate to the risks at low doses and low-dose rates.
5. The transfer of risk estimates based on a Japanese population to other populations with quite different natural incidences of many cancers.

The reduction of these uncertainties requires further studies and a greater understanding of the effects of radiation which will entail both epidemiological and experimental approaches. The correct methods of the projection of risk over time, and the transfer of risk across populations are not known. In the case of the former, the answer will become evident as the epidemiological studies progress in time. In the case of transfer of risk across populations, epidemiological studies of the risks for the induction of cancer at the relevant sites in populations other than the Japanese are needed.

How to extrapolate from data obtained for exposures to high doses and high-dose rates to the risks incurred at low doses and low-dose rates is, perhaps, more likely to come from understanding the processes involved in radiation-induced cancer than from epidemiological studies.

This Report gives an account of the progress that has been made in the last decade or so in the understanding of many of the facets of the difficult task of estimating the risk of events of low probability after low levels of irradiation. At the same time, the better understanding that comes with progress has made it possible to identify more clearly the gaps in the knowledge of how best to estimate risks of the effects of radiation. These risks are so influenced by age at exposure, gender, inherent susceptibility, which is influenced not only by genetic background, but by diet and exposure to various agents, that it makes it evident that many of the questions to be answered are very complex. However, answers to the problem of how to extrapolate from high doses and dose rates to low doses and dose rates, how to project estimates of risk over time, and how to transfer risk estimates from one population to another, would go a long way in the improvement of current estimates. Finally, it should be noted that the current estimates provide a robust basis for radiation protection recommendations.

2. Introduction

In 1987, the NCRP published Report No. 91, *Recommendations on Limits of Exposure to Ionizing Radiation* (NCRP, 1987). That report was an important step forward in radiation protection recommendations and introduced both substantial revisions and novel concepts, including the cumulative exposure guidance level for occupational exposure, remedial action levels for natural and other sources of radiation, and a negligible individual risk level for defining sources for which further effort to reduce radiation exposure to the individual was deemed to be unwarranted.

In 1987, the available estimates of risk of radiation-induced cancer were a decade old (ICRP, 1977; NAS/NRC, 1980; UNSCEAR, 1977), although it was emphasized in Report No. 91 (NCRP, 1987) that it was known that forthcoming cancer-risk estimates, based on new information from Japan, would be higher. Some of the provisions in the report, such as the cumulative guidance, were based on an awareness of increased risk estimates. At that time, the degree of increase in the risks was the subject of international discussions which have now reached fruition (ICRP, 1991; NAS/NRC, 1990; UNSCEAR, 1988).

Rapid developments have taken place in the evaluation of cancer-risk estimates since 1987. The evaluation of the effects in the survivors of Hiroshima and Nagasaki has been especially important in view of the revisions in the dosimetry for the survivors, the accumulation of 11 more years of data for solid tumors, improvements in statistical methods and in the data base for risk projections (Shimizu et al., 1987; 1989). Definitive reports on these developments have been published by the Radiation Effects Research Foundation (RERF) in Japan, for example, on the changes in the dosimetry (Roesch, 1987) and its impact (Preston and Pierce, 1988) culminating in a general epidemiological review of the data (Shimizu et al., 1988). Table 2.1 provides a listing of the total number of malignancies in the survivor population by site as well as an estimate of the excess number of malignancies by site for all dose categories, all ages at exposure and both sexes. Considering the small numbers of excess cancers, it is not yet possible to apply models for projection to lifetime risks other than those that are relatively simple models.

7

TABLE 2.1—*Atomic-bomb survivor data for the period of 1950 to 1985.*

Site of cancer	Total cancer Cases[a]	Number of excess Cancer cases[b,c,d]
Leukemia	202	78
All cancers except leukemia	5,734	266
Esophagus	176	11
Stomach	2,007	72
Colon	232	19
Lung	638	44
Female breast	155	22
Ovary	82	10
Urinary tract	133	19
Multiple myeloma	36	8
Remainder	2,275	61
Total for all sites	5,936[e]	344[f]

[a]Data from Shimizu *et al.* (1988).

[b]Assumes an average shielded kerma of 0.162 Gy.

[c]The number of cases are for all exposure categories up to and including the 4 + Gy category.

[d]The equation used to calculate the number of excess cancers by site is:

$$E = \frac{O \times R \times D}{1 + (R \times D)}$$

where E is the excess fatalities, O is the observed cancer fatalities, R is the excess relative risk of fatality by cancer site and D is the dose in average shielded kerma.

[e]Based on a subcohort of about 76,000 survivors.

[f]This estimate may be compared to the estimate for 1950 to 1974 of 190 (Beebe *et al.*, 1978a) and 251 for 1950 to 1978 (Kato and Schull, 1982).

General evaluations of all sources of cancer risk, emphasizing especially the data from Hiroshima and Nagasaki, have been provided by UNSCEAR (1988), BEIR V (NAS/NRC, 1990) and NRC (1991). These evaluations provide a wealth of risk information on the carcinogenic and genetic effects of radiation, but they stop short of providing the necessary definitive judgments and indeed, in some cases, the precise data, to establish an adequate basis for estimates of risk (or detriment) at low doses for use in making radiation protection recommendations. To cite just one example, it is necessary to proceed from the UNSCEAR (1988) and BEIR V (NAS/NRC, 1990) high dose, high-dose-rate data, for which risk estimates are available, to the low dose, low-dose rate or very low doses at high-dose-rate situations pertinent to radiation protection. To estimate the effects of low doses and low-dose-rate irradiation, UNSCEAR (1988) advises using a reduction factor of "from two to ten," and BEIR V (NAS/NRC, 1990) suggests using "two or more." Neither report recommends a specific factor to obtain low-dose estimates. Consequently, it is necessary for radiation protection bodies, such as the NCRP, to consider all the new information available in the UNSCEAR (1988) and BEIR V

(NAS/NRC, 1990) Reports and from other sources and then to provide specific recommendations on the most appropriate risk estimates to be used for radiation protection in the low-dose range. The ICRP (1991) based its estimate of lifetime risk of excess mortality for radiation-induced cancer on the estimates in the UNSCEAR Report (UNSCEAR, 1988) and the BEIR V Report (NAS/NRC, 1990), utilizing a dose and dose-rate effectiveness factor (DDREF) of two.[1]

The purposes of this Report, therefore, are to examine:

1. The data provided in the UNSCEAR (1988) and BEIR V (NAS/NRC, 1990) Reports, noting both similarities and differences from each other and from previous reports and to assess their completeness and adequacy
2. The methods of analysis including dose-response relationships, risk-projection models and derivations of risk at low dose and dose rate from data obtained at high doses and dose rate
3. Other sources of data that may be needed and thus provide:
   a. what other data may be required to estimate the risk of cancer and other late effects from exposures to low doses and/or low-dose rates in both the general and worker populations in the United States for the purpose of radiation protection
   b. the distribution of organ risks for use in providing organ dose weighting factors

This information is intended to provide the primary bases for the NCRP's basic recommendations, *Limitation of Exposure to Ionizing Radiation* (NCRP, 1993).

---

[1]NCRP Report No. 64 (NCRP, 1980) introduced the term dose-rate effectiveness factor (DREF) and ICRP Publication 60 (ICRP, 1991) used dose and dose-rate effectiveness factor (DDREF). Both abbreviations are used in this Report and they are considered to be synonymous.

3. United Nations Scientific Committee on the Effects of Atomic Radiation Reports

3.1 Earlier Reports of the United Nations Scientific Committee on the Effects of Atomic Radiation

There have now been ten reports to the General Assembly by UNSCEAR since the committee was established in 1955, although not all of these reports present risk estimates for induced cancer. In the earlier reports, beginning in 1958, the estimates of radiation-induced cancer were restricted to leukemia because of the lack of data for solid cancers. The evolution of UNSCEAR's risk estimates up to 1977 is shown in Table 3.1.

In the 1977 UNSCEAR Report, the section on carcinogenic effects of radiation was introduced with the statement that it has become clear that the most important late somatic effect of low doses of radiation is the infrequent induction of malignant disease. For many years, after Müller's discovery of the mutagenic effect of radiation (Müller, 1927), there was more concern about genetic effects than about cancer. This is not the situation today, however. The consensus about the predominant importance of the carcinogenic effect developed as a result of the epidemiological studies of radiation-induced cancer. The findings of the study of the atomic-bomb survivors had a major impact in changing the primary concern from genetic effects to cancer induction. The 1977 UNSCEAR Report was the first to provide risk estimates for cancer for a range of organs.

In the UNSCEAR Report (UNSCEAR, 1977), the data from various exposed populations were considered and were used in the risk estimates for cancer in specific organs. The principal source of data used for the estimate of the risk of radiation-induced excess cancer was the atomic-bomb survivors. The so-called Tentative 1965 Dosimetry Reconstruction (T65DR) dosimetry was used; a salient feature was that 20 percent of the total absorbed dose at Hiroshima was considered to be contributed by neutrons. For examination of dose-response

10

relationships, the population was stratified among various dose groups that were arbitrarily divided into 0.1 to 0.49, 0.5 to 0.99, 1 to 1.99 and >2 Gy for most of the analyses.

In 1977, estimates of risk of excess cancer were made for many more sites than previously (Table 3.1). The risk coefficients were highest for thyroid, lung and breast cancer in females, and for leukemia. A distinction between mortality and incidence rates was made and in the case of the thyroid, for example, the difference was large. The rate of induction of thyroid cancer in some populations was about 1×10^{-2} Gy^{-1} with a mortality rate about tenfold less.

It can be seen from Table 3.1 that the risk of mortality was estimated for only eight specific sites, because it was not possible to estimate directly with sufficient accuracy either risks for other specific cancers or the total risk of excess cancer from the atomic-bomb survivors. It was realized that it was likely that most of the solid

TABLE 3.1—*Summary of UNSCEAR estimates of cancer-risk coefficients (percent Sv^{-1}) (adapted from Table 3, UNSCEAR, 1988).*[a]

Tissue	1958 Report	1964 Report	1972 Report	1977 Report
Bone marrow	0.2–0.5	0.01–0.02[b]	0.14–0.40	0.20–0.50
Breast			0.06–0.20[c]	0.50
Lung			0.0006–0.02[d]	0.25–0.50
Thyroid		0.16	0.20–0.80[e]	0.10
Stomach				0.10–0.15
Liver				0.10–0.15
Brain				0.10–0.15
Large intestine				0.10–0.15
Salivary gland				0.10–0.15[f]
Small intestine				
Bone				
Esophagus			0.40	
Bladder				
Pancreas				0.02–0.05[f]
Rectum				
Mucosa of cranial sinuses				
Lymphatic tissue				
Skin				low
Estimated total				1.0–2.5

[a]The risk estimates in the reports of 1958 to 1977 were based on the rad as the unit of dose. These have been changed to Sv. Mortality rates are used except as indicated.

[b]Risk per year.

[c]Excess mortality for 25 y period.

[d]Excess mortality per year. The estimate of incidence over 25 y was 0.12 to 0.4 percent per Sv.

[e]Incidence over 20 years. The incidence in females was about twice that of males.

[f]Incidence.

cancers had not yet occurred and the risk estimates for these cancers were imprecise. Because sufficient data for solid cancers were not available, it was necessary to estimate the total lifetime excess cancer risk based on an estimate of the ratio of the lifetime estimate of cancer risk including leukemia to the estimate for the lifetime risk of leukemia, *i.e.*, a ratio of approximately five.

The estimate of the total lifetime excess cancer risk was made in the following way. The risk of excess leukemia was estimated by pooling acute and chronic leukemias and using organ doses based on T65DR that took into account the RBE for neutrons. The lifetime risk for leukemia was estimated to be about 0.5×10^{-2} Gy^{-1} for doses of a few Gy and 0.2×10^{-2} Gy^{-1} for doses in the range of 1 Gy. The ratio between the frequency of solid cancers and of leukemias was taken to be four to six, implying that the total excess cancer rate would be about 2.5×10^{-2} Gy^{-1} for the higher doses and 1×10^{-2} Gy^{-1} for lower doses. The choice of a ratio of four to six was a judgment derived from the study of the patients with ankylosing spondylitis at the time. The risk estimates for leukemia were derived mainly from data obtained at doses greater than 1 Gy. The following caveat was therefore added to the higher risk estimate: "While the rate per unit dose from doses of a few rad is unlikely to be higher than this value it might be substantially lower." It is this uncertainty that remains today and is central to the estimates of risk of excess cancer as a result of occupational or environmental radiation exposure.

The 1977 UNSCEAR Report noted the influence of age and sex, the young and females being more susceptible than adults or males, respectively. The special case of the fetus was considered, and it was noted that even at low doses the lifetime risk of fatal malignancies for exposures during gestation might be of the order of 2 to 2.5×10^{-2} Gy^{-1}.

3.2 The 1988 Report of the United Nations Scientific Committee on the Effects of Atomic Radiation

The 1988 UNSCEAR Report is the tenth of the comprehensive series documenting the increasing knowledge of radiation exposure and the associated acute and late effects, particularly cancer. In the 1988 UNSCEAR Report, excess cancer mortality was taken to be the major concern and the major factor to be used in setting protection standards.

3.2.1   *Sources of Data*

In the UNSCEAR 1988 Report, the major changes from previous reports in estimates of the risk of health effects concern cancer mortality. The committee had the advantage of having at hand the RERF reports and the corresponding papers, both published and in press; in particular, Preston and Pierce (1988) and Shimizu *et al.* (1987; 1988; 1989). Cancer mortality had been analyzed among 75,991 atomic-bomb survivors of the Life Span Study (LSS) in the period 1950 to 1985, for whom doses based on the DS86 (Roesch, 1987) were available. There is a total of 120,128 persons in the LSS, 26,517 of whom were not exposed (not in city) and 15,237 persons for which T65DR doses were known but for whom DS86 doses were not available in 1987. There were 2,383 persons exposed for whom a dose (T65DR) could not be estimated. The analyses are based on a total of 5,936 cancer deaths. Although 1985 was 40 y after the bombing, over 60 percent of the survivors were still alive, and in particular, those who were young ATB. It is thought that those exposed at a young age now appear to be at greater risk of radiation-induced cancer than those exposed at older ages.

As in previous reports, the UNSCEAR Report (UNSCEAR, 1988) also draws upon data obtained from other exposed populations, particularly the patients with ankylosing spondylitis treated with acute partial-body exposures to x rays (Darby *et al.*, 1985; 1987) and patients with cervical cancer treated with x or gamma rays (Boice *et al.*, 1987).

Follow-up of the 14,106 ankylosing spondylitic patients is as long as 48 y for some patients. Perhaps the most interesting finding, in the most recent study of the ankylosing spondylitic patients, is the apparent decrease in the relative risk of solid cancers about 25 y after exposure. If such a decrease were a general finding, it would influence the choice of the risk projection model. While the populations, particularly the cervical cancer population, are large, they are not representative populations, because of age distribution, sex or clinical status. Also, these populations experienced partial-body irradiation with high doses to the treated areas. Therefore, the dosimetry and the relationship of the risks derived from these studies to those after whole-body irradiation is complex. However, data from such studies are important for purposes of confirmation of risk estimates. The study of the atomic-bomb survivors is the only study of the effects of whole-body irradiation of a large population consisting of people of all ages, both sexes and for a broad range of doses.

Between 1977 and 1988 both of the major components of risk assessment, namely, cancer mortality and estimated doses for the atomic-bomb survivors changed. In the decade of follow-up of the atomic-bomb survivors after 1975, the deaths from cancer considered to be associated with the radiation almost doubled, although the number is still not large, namely about 344 of the 5,936 cancer deaths reported (Pierce, 1989). The increase is largely, but not entirely, due to excess solid cancers. Despite the additional data, cancer-risk estimates for mortality were made for only eight specific organ sites and for leukemia. The estimates of radiation-induced leukemia were still based on data pooled from different types of leukemia, excluding chronic lymphocytic leukemia.

While more information has accumulated about the experience of a number of different human populations exposed to radiation, the atomic-bomb survivors are the largest population that has experienced a single brief whole-body exposure. Their doses are relatively well known and their fate is known with considerable precision. Thus, this population remains the major source of data for risk estimation for exposures to low-LET radiation.

3.2.2 *Dosimetry*

The most recent assessment of the atomic-bomb radiation dosimetry, *i.e.*, DS86 was published in 1987 (Roesch, 1987). As a result of this reassessment, the estimated yield of the Hiroshima bomb was increased from 12 ± 2.5 kt to 15 ± 3 kt. Newly calculated spectra and transport codes resulted in the free-in-air gamma kerma in Hiroshima being increased by about a factor of three at large distances from the hypocenter as compared to the earlier T65DR. The free-in-air neutron kerma was decreased to about ten percent of the T65DR estimate. The yield of the Nagasaki bomb was adjusted down to 21 ± 2 kt from 22 ± 2 kt, and based on the new calculations, the DS86 free-in-air gamma kerma estimate was somewhat less than that based on T65DR (Figure 3.1).

The DS86 provides free-in-air kerma, shielded kerma and organ doses for a high proportion of the individuals exposed. Not only is location in relation to the hypocenter based on coordinates on the map taken into account, but also orientation, posture, location in the open or in a house, and body size based on the age of the survivor ATB. When the DS86 is applied, it alters the distribution of the survivors in the various dose groups from those existing for the former T65DR, thus precluding the use of a simple change in the dose scale.

Organ doses, the requisite for risk estimates, were obtained by first calculating the free-in-air kerma and then estimating attenuation by

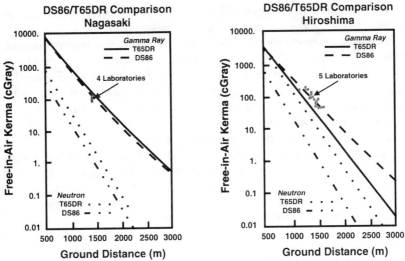

Fig. 3.1.   Comparison of DS86 and T65DR free-in-air kerma: Nagasaki and Hiroshima (from Sinclair and Preston, 1987).

the walls and roofs of houses and buildings to obtain the shielded kerma. Finally, taking into account body size (based on age) and position, doses for 15 organs were obtained by correcting for the attenuation of the radiation in the body on an individual basis. The estimated attenuation of gamma rays by the structures in the DS86 is about twice that of T65DR for both cities. There is very little difference in the estimate of the neutron kerma attenuation between the two dosimetry systems in either city. In contrast, the transmission of the gamma rays through the body is now estimated to be greater than previously calculated. The increases in gamma-ray-kerma attenuation due to shielding and the greater transmission through the tissues of the body tend to cancel each other out. So, although there has been a change in the estimates of the various contributors to the dose, the actual change in organ doses depends mainly on the reduction of the neutron contribution and on the value of the RBE applied, and is not greater than a factor of two different from previous estimates.

At the time of the analysis by UNSCEAR (1988), dose calculations were not yet completed for about 18 percent of the survivors, most of whom were shielded by the terrain or were in concrete buildings, some of which contained heavy equipment, which made the determination of the shielding very difficult. This segment of the population in Nagasaki is of particular importance because it constitutes a significant fraction of the survivors in Nagasaki with estimates of free-in-air kerma between 0.5 and 2.0 Gy. Recently, doses for an additional 11,000 survivors have been calculated.

The new dosimetry system has made use of many improvements, especially in methods of computation and in verification, resulting in a system in which there is considerable confidence. Especially important is the experimental verification of gamma-ray doses by thermoluminescent measurement techniques over virtually the full range of survivor doses. The confidence might be complete if the apparent paradox of the existence of survivors for whom doses are estimated to exceed 6 Gy did not still exist. Estimates of the errors in the dosimetry have also been improved. The uncertainties are often considered to be of two main types. First, systemic errors, which are the same for each survivor, and second, random errors, which apply independently to each survivor.

Central to the accuracy of the dosimetry is the accuracy of the assignment of the location, orientation in relation to shielding and posture of the individual survivors. The accuracy of these important factors depends on the memory of the survivor. Of comparable importance are such items as the yield of the bombs, the transport through air of the radiation, which depends on such factors as air density and moisture, topography, shielding and calculation of organ doses.

There has been a number of attempts to use reports of biological effects in survivors, such as epilation and effects on the gastrointestinal tract (Gilbert and Ohara, 1984; Jablon, 1971; Stram and Mizuno, 1989), to evaluate and compare T65DR and DS86. Stram and Mizuno were unable to establish that DS86 was a better predictor of epilation response than T65DR. However, among the various biological studies, it is now clear that there are mostly smaller differences between Hiroshima and Nagasaki than formerly estimated. Indeed, for fatal cancer, differences between the two cities still exist but they are no longer as significant. This, too, gives greater confidence in DS86 than T65DR. Nevertheless, DS86 should not be considered the final word. Agreement between the two cities is much better, but not perfect. The measured thermal neutron fluxes at Hiroshima exceed calculated values at further distance (Roesch, 1987) and this unresolved problem casts some doubt on our knowledge of the fast neutron versus distance curve at Hiroshima. There is still doubt about the best estimate of the size of the neutron component with distance and its energy spectrum and they are still under investigation. In addition, since the risk estimates are derived largely from total doses in excess of 1 Gy, where the RBE for the associated neutron doses may be falling with increasing dose from its maximum value, and the deviation between calculated and measured fluxes is not large, the impact of any revision of the neutron doses will be relatively small. Any increase in the estimate of neutrons will, of course, reduce the risk estimates, but it is considered unlikely that any decrease will be greater than 25 percent (Preston *et al.*, 1992–1993).

The contribution of delayed radiation from residual radioactivity produced by neutron activation is greater in the DS86 than T65DR. There are sources of error in the estimates of the delayed radiation as well as for the prompt radiation. For example, the effect of the blast and other factors no doubt changed the subsequent shielding of many of the survivors, estimates of the absorbed dose from delayed radiation still involve some error. The total uncertainty is estimated to be between 30 to 45 percent.

3.2.3   *Data Analysis*

In the analysis of Shimizu *et al.* (1988), the major source for the UNSCEAR 1988 estimates of cancer risks, person years and the number of cancer deaths were aggregated and stratified by city, sex, age ATB, follow-up period or attained age, and dose. The ages ATB were grouped <10, 10 to 19, 20 to 29, 30 to 39, 40 to 49 and >50 y of age, and attained ages <20, 20 to 29, 30 to 39, 40 to 49, 50 to 59, 60 to 69 and >70 y with follow-up periods of 5 y intervals from 1950 through 1985. The dose categories were 0, 0.01 to 0.05, 0.06 to 0.09, 0.1 to 0.19, 0.2 to 0.49, 0.5 to 0.99, 1.0 to 1.99, 2.0 to 2.99, 3.0 to 3.99 and >4.0 Gy.

The excess cancer-risk coefficients (Table 3.2), adopted by UNSCEAR (1988) are those obtained by Shimizu *et al.* (1987; 1988) which were averaged over cities, sexes and ages ATB. The coefficients are the mean values for both cities, both sexes (with the exception of the ovary and the breast) and all ages ATB for leukemia, excluding chronic lymphocytic leukemia, all cancers except leukemia, female breast, lung, esophagus, stomach, large intestine (excluding rectum), ovary, bladder and multiple myeloma. The coefficients of Shimizu *et al.* (1987; 1988) were derived using the DS86 absorbed organ doses, a linear-dose response, a constant relative risk model for all cancers except leukemia, an absolute risk model for leukemia and an assumption of an RBE of one. The cancer sites were restricted to those sites for which the mortality increased significantly with dose. The analyses by Shimizu *et al.* (1987; 1988) were based on cancer mortality up to 1985. In 1985, some 40 y after the exposure to the atom bomb radiation, over 60 percent of the exposed population were still alive.

Significant fractions of other irradiated populations are also still alive, therefore it is necessary to use risk-projection models to project

TABLE 3.2—*Comparison of lifetime risk estimates by (UNSCEAR, 1977; 1988). (Estimates of excess fatal cancer per 10,000 individuals, for all ages and both sexes at 1 Gy, by tissue).*[a]

	1977[b,c]	1988[d] Additive Model	1988[d] Multiplicative Model
Bone marrow	50	93 (77–110)	97 (70–132)
Bladder	5	23 (11–40)	39 (16–73)
Breast	50	43 (22–69)	60 (28–105)
Colon		29 (14–46)	79 (36–134)
(intestines & rectum)	15		
Esophagus	5	16 (3–31)	34 (8–72)
Lung	50	59 (34–88)	150 (84–230)
Multiple myeloma		9 (3–17)	22 (6–51)
Ovary		26 (8–48)	31 (9–68)
Stomach	15	86 (45–131)	130 (66–199)
Remainder	60	103	114
Total	250[e]	450[e]	710[e]
All cancers except leukemia		360 (280–440)	610 (480–750)

[a]RBE = 1.

[b]In the estimation in 1977 the unit of dose was the rad.

[c]For each organ the highest risk value given in the range of risks in UNSCEAR (1977) is shown.

[d]For each organ the mean risk estimate and the 90 percent confidence limits (in parentheses) are shown.

[e]Age-averaged risk coefficients.

estimated lifetime risks. A minimum latency period of 2 y for leukemia and 10 y for all other cancer sites was assumed. The period of excess risk was assumed to be 40 y for leukemia and lifetime for other cancers. Estimates were made for the total population, the population over 25 y of age and a worker population, 25 to 64 y of age. Estimates were computed separately using constant additive and constant multiplicative risk projection models. Estimates of years of life lost due to excess cancer mortality were also calculated. For individual organ risks, age-average risk coefficients were used.

The coefficients for excess cancer mortality were estimated using a parametric model developed by the Centre d'Etude sur l'Evaluation de la Protection dans le Domaine Nucleaire. This model, which uses standard lifetable techniques, permits several kinds of computations of the effects on cohorts of specific age and sex as well as on populations with different age distributions.

The method uses a doubly decremented lifetable in which the age-specific death rate for any given exposure history is the sum of the age-specific mortality rate in a nonexposed population and radiation-related excess cancer rate estimated for that age and exposure history.

The 1982 Japanese population served as the reference population for the calculation of lifetime risks based on risk coefficients obtained from the atomic-bomb survivors. The age distribution of the 1982 Japanese population was used for projection of risk. Equal numbers of both sexes were used for estimating risks for both the worker population and the whole population. The current adult male and female populations in the United Kingdom were used as the reference population for the ankylosing spondylitic patients and the female population in the United States for the cervical cancer studies.

The 1988 UNSCEAR lifetime-cancer-risk estimates for specific organs in the general population, based on the atomic-bomb survivors and mortality data for the 1982 Japanese population, are shown in Table 3.2. The 1977 UNSCEAR estimates are shown for comparison.

It can be seen that the estimated excess cancer-risk for all sites is 7.1 and 4.5 percent Sv^{-1} using the multiplicative and additive risk projection models, respectively. These estimates can be compared with the 1977 estimate for high doses of about 2.5 percent Sv^{-1} based on an additive model. However, the estimates shown in Table 3.2 were based on age-averaged risk coefficients and better values for the estimation of total risks are obtained from age-specific-risk coefficients. The corresponding estimates based on age-specific coefficients are 4 and 11 percent Sv^{-1} based on the additive and multiplicative projective models, respectively (see Table 3.3).

The number of person years of life lost from all types of cancer in a working population is shown in Table 3.4. The divergence in lifespan lost between the additive and multiplicative projection models is smaller than the divergence of fatal cancer deaths given in Table 3.3 for the same risk coefficients. However, the loss of life expectancy is greater for the additive model than for the multiplicative model.

TABLE 3.3—*Projections of lifetime probability of fatal cancer for 10,000 persons (5,000 males and 5,000 females) exposed acutely to 1 Gy whole-body low-LET radiation (adapted from Table 71, Annex F, UNSCEAR, 1988).*

	Projection model	Excess fatal cancers[a]	
Total population	Additive	400[b]	500[c]
	Multiplicative	1,100[b]	700[c]
Adult population[d]	Additive	—	500[c]
	Multiplicative	—	600[c]
Worker population	Additive	600[b]	400[c]
(ages 25 to 64 y)	Multiplicative	700[b]	800[c]

[a]Rounded values.
[b]Based on age-specific coefficients of probability.
[c]Based on age-averaged coefficients of probability.
[d]Age groups 20 to 29, 30 to 39 and 40 + y of age.

TABLE 3.4—*Estimates of loss of life expectancy (person years) in 1,000 male or 1,000 female adult atomic-bomb survivors exposed to 1 Gy organ dose of high-dose-rate, low-LET radiation[a] (from Table 60, Annex F, UNSCEAR, 1988).*

Malignancy	Sex	Risk projection model	
		Additive	Multiplicative
Leukemia[b]	M	290	140
	F	170	120
All other cancers[c]	M	500	400
	F	700	530
Total (average)		830	600

[a]Using age-averaged risk coefficients for all age groups >25 y for the Japanese 1982 population.

[b]Assumes a plateau of 40 y for leukemia.

[c]Assumes a lifetime plateau for all other cancers.

In the additive model, excess risk is spread out evenly over the expression period, whereas in the multiplicative model it varies in direct proportion to the underlying age-specific population cancer rate. Thus, according to the additive model, most of the cancers occur at relatively young ages, when there are more people at risk, whereas according to the multiplicative model, they occur at relatively late ages when population rates are high. Because the risk coefficients for exposure at young ages are usually based on limited follow-up, mainly at young ages, estimated lifetime excess risk based on the additive risk projection model for any exposed population that includes a substantial proportion exposed at young ages will be less than the corresponding multiplicative model estimate. But those fewer cancers will be predicted to occur mainly at younger ages. As the follow-up time is increased, the estimates of lifetime risk according to the two models must converge, but the estimates of loss of life expectancy will diverge, until the absolute model estimate is considerably greater than the multiplicative model estimate.

The excess cancer mortality for some sites was significant (at the 95 percent confidence limit) at doses between 0.2 and 0.5 Gy. There was no significant increase in cancer mortality below 0.2 Gy of organ dose. This is the first time the data suggest a statistically significant excess for some specific cancers in the atomic-bomb survivors, *e.g.*, bladder cancer.

3.2.4 *Influence of Age at Exposure*

There is a suggestion that the risk of leukemia after exposure to ionizing radiation is higher in those exposed when young, especially in the first decade of life, than in those exposed at older ages (see Table 3.5).

The data for leukemia in the atomic-bomb survivors indicate that the influence of age at the time of exposure is complex (Figure 3.2). Susceptibility is high in early childhood, decreases rapidly reaching a low level in the teens but rises again in middle age. The pattern of risk of leukemia in relation to age at exposure is probably confounded by the fact that different types of leukemia have been pooled. In the case of the female breast, susceptibility to cancer induction by radiation exposure appears to be higher in childhood and the risk decreases markedly later in life (Tokunaga et al., 1987), Figure 3.3. The lifetime risk estimates for cancer in those exposed early in life have large errors because of the small numbers of persons in the cohorts. It will be some years before the relationship of age at exposure and risk of radiation-induced solid cancers can be determined definitively because those exposed at a young age have only started to reach the age of expression of cancer risk.

UNSCEAR (1988) compared the projected lifetime risk estimates between the total Japanese population, adult population and a so-called worker population with an age range of 25 to 64 y (Table 3.3). The inclusion of the young in the population increases the total risk when a multiplicative risk-projection model is used. These

TABLE 3.5—*Excess deaths per 10^4 person-y Gy by age ATB and age at death based on DS86 shielded kerma (from Shimizu et al., 1988).*[a]

Age ATB	Total	Age at death						
		<20	20–29	30–39	40–49	50–59	60–69	>70
Leukemia								
<10	2.93	6.71	0.93	1.27	−0.01			
10–19	1.19	3.95		0.56	0.02	−0.06		
20–29	2.13		3.93	1.52	4.84	0.01	−0.28	
30–39	2.54			0	3.18	2.26	1.09	3.89
40–49	2.11				−0.35	3.07	−0.24	3.50
>50	4.56					4.31	3.84	5.12
Total	2.29	6.48	2.17	1.16	1.88	1.54	1.09	4.24
All cancers								
Except leukemia								
<10	2.29	(0.43)	1.32	2.85	5.16			
10–19	4.66	(3.96)	(−0.12)	2.00	5.84	13.91		
20–29	9.38			(1.39)	9.40	15.71	14.33	
30–39	9.31			(−1.32)	(1.33)	3.16	11.00	41.01
40–49	14.52				(2.48)	(3.37)	7.31	37.30
>50	7.89					(35.29)	(−2.88)	17.21
Total	7.41	0.79	0.54	1.98	5.35	9.62	6.85	30.53

[a]Numbers in parentheses are excess deaths before the assumed minimum latent period of ten years.

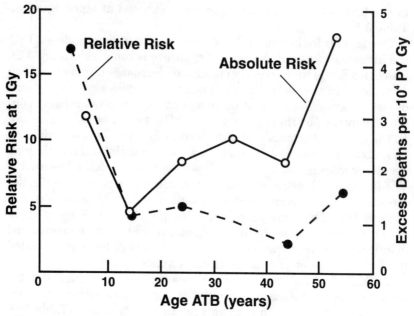

Fig. 3.2. Relative and absolute risks of leukemia, all types except chronic lympho-cytic leukemia in atomic-bomb survivors (1950 to 1985) as a function of age at exposure (from Upton, 1991).

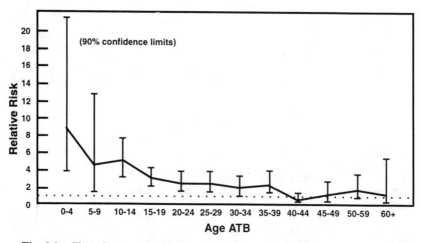

Fig. 3.3. The relative risk of breast cancer in atomic-bomb survivors exposed to ≥ 0.5 Gy as a function of age ATB. The atomic-bomb survivors exposed to 0 to 0.09 Gy kerma, both in and "not in city," constituted the control group. The 90 percent confidence intervals are indicated (from Tokunaga *et al.*, 1987).

categorizations of risks by age range are helpful in establishing exposure guidelines for various age groups.

3.2.5 Influence of Sex on Risk Estimates

The influence of sex on the risk estimates noted in the 1977 UNSCEAR Report are seen again in UNSCEAR (1988), but the differences in risk for males and females are not as great as estimated in a report focused on radioepidemiological tables (NIH, 1985). The excess mortality for solid cancers in the female atomic-bomb survivors is greater than in the male survivors (Table 3.6), but less for leukemias based on absolute risks.

The sex-dependent difference in the estimates of risk of leukemia is influenced by the choice of the projection model. The risk based on the additive risk-projection model is nearly twice as high in males as females; but with multiplicative projection model the difference is about ten percent.

For all other cancers, the risk is about 40 percent higher for females than for males with the additive risk projection model assuming the duration of the plateau is lifetime. With the multiplicative projection model the risk for females is approximately 25 percent higher than it is for males, see Table 3.6. The sex-dependent difference in the risk of total cancer is about the same as was assumed in ICRP Publication 26 (ICRP, 1977) which was about 50 percent. In the analysis by Land and Sinclair (1991) it was found that the sex-dependent difference decreased with age, from about 50 percent in the young to about ten percent at older ages.

TABLE 3.6—*Excess cumulative lifetime cancer mortality from 1 Gy acute low-LET radiation in adulthood[a] (from UNSCEAR, 1988).*

Type of cancer	Sex	Deaths per 10^4 persons	
		Additive risk[b] Projection Model	Multiplicative risk[b] Projection model
Leukemia	M	130[c]	90[c]
	F	70[c]	81[c]
All other cancers	M	300[d]	410[d]
	F	420[d]	520[d]

[a] ≥ 25 y old.
[b] Age-averaged coefficient.
[c] Duration of plateau is 40 years.
[d] Duration of plateau is lifetime.

3.2.6 Comparison of Risk Among Irradiated Populations

There are many epidemiological studies of the effect of radiation on cancer induction at specific sites. Most of the populations studied, as noted earlier, have features such as age distribution, dose distribution, sex differences and health status that restrict the use of the risk estimates. There are two major studies that were considered in UNSCEAR (1988) as suitable for comparison with the atomic-bomb survivor study, at least to establish whether large discrepancies in risk estimates existed. It should be noted that in the two populations selected, therapeutically irradiated patients with ankylosing spondylitis (Darby et al., 1985; 1987) and the cervical cancer patients (Boice et al., 1985; 1987), the radiation exposures were quite different, particularly in dose distribution, both from each other and from that experienced in Hiroshima and Nagasaki. The ankylosing spondylitis patients were treated with partial-body exposures to x rays. The radiation treatment of the cervical cancer patients was given by external beam, commonly ^{60}Co gamma rays or by intracavitary application of radium, or a combination of both. The populations in Nagasaki and Hiroshima experienced a single exposure to gamma rays at a high-dose rate and, in the case of those survivors near to the hypocenter, to a small but possibly significant level of neutrons.

The comparison of the projected risks in the three populations is shown in Table 3.7.

The risk estimates for the ankylosing spondylitis patients are comparable in the case of leukemia to those for the atomic-bomb survivors but considerably less for solid cancers. Given the

TABLE 3.7—*Projection of excess lifetime cancer mortality from adult population[a] of 1,000 males or 1,000 females exposed to 1 Gy low-LET radiation at high-dose rate[b] (from Table 59, Annex F, UNSCEAR, 1988).*

Malignancy	Sex	Atomic-bomb survivors (Japan)[c]		Ankylosing Spondylitis (UK)[d]		Cervical cancer Patients (Multinational)	
		Additive	Multiplicative	Additive	Multiplicative	Additive	Multiplicative
Leukemia[e]	M	13.0	9.0	4.4	14.0	—	—
	F	7.0	8.1	—	—	1.4	2.8
All other cancers[f]	M	30.0	41.0	7.8	23.0	—	—
	F	42.0	52.0	—	—	—	—

[a] Over 25 y of age.
[b] Doses received within seconds or minutes.
[c] Reference population: Japan, 1982.
[d] Reference population: United Kingdom, 1982.
[e] Plateau is 40 years.
[f] Plateau is lifetime.

differences in the populations at risk and the variable length of follow-up for the individual patients in the ankylosing spondylitis and cervical cancer series, it is difficult to evaluate the consistency of risk estimates among the atomic-bomb survivors in Japan, the ankylosing spondylitis patients in the United Kingdom and cervical cancer patients in the United States and other countries.

The number of leukemia cases is not expected to increase markedly with increased follow-up in any of the populations under study. Therefore, the projected risk estimates of leukemia are unlikely to change markedly. In contrast, the number of solid cancers is likely to increase in the atomic-bomb survivors, and this potential increase is taken into account in the projection of risk to lifetime risks.

3.2.7 Projection of Risks for Different Populations

3.2.7.1 Young Age Groups. Based on organ doses, comparisons were made in UNSCEAR (1988) among average risks over the period 1950 to 1985 for age groups 0 to 9 ATB, 10 to 19 ATB and all age groups together (Table 3.8), for both absolute risk and excess relative risk, as reported by Shimizu *et al.* (1988). It is of interest that the absolute and relative risks for leukemia in the 10 to 19 age group are smaller than those for a population of all ages, but for solid cancer the relative risk is higher. This may reflect the differences in both the biology and radiation responses of childhood and adult leukemias and lymphomas or it may be associated with the small number involved. This finding may also indicate the limitations of pooling data of quite distinct types of leukemia and other hematological tumors for analysis. Subdivisions into specific types would be worthwhile, as indeed was done in earlier studies of the Japanese data.

3.2.7.2 Adults. The lifetime projections of excess mortality for solid cancers for atomic-bomb survivors are greater than those for

TABLE 3.8—*The absolute and excess relative risk of cancer in relation to age.*

Age group	Leukemia		All cancer excluding leukemia	
	Absolute Risk[a]	Excess relative Risk[b]	Absolute Risk[a]	Excess relative Risk[b]
0–9[c]	3.4	19.1	2.8	1.6
10–19[c]	1.5	4.5	6.2	1.0
All ages[d]	2.9	5.2	10.1	0.4

[a]Excess mortality per 10^4 person-y Gy.
[b]Excess relative risk at 1 Gy organ absorbed dose.
[c]From Table 5A, Shimizu *et al.* (1988); Table V, Shimizu *et al.* (1990).
[d]From Table 54, Annex F, UNSCEAR (1988).

the ankylosing spondylitis patients (Darby *et al.*, 1985) using the additive or multiplicative projections reflecting the larger risk coefficients (see Table 3.7). The regimen of exposure and many other factors contribute to the expectation that absolute risk in the ankylosing spondylitis study would be different from that for the atomic-bomb survivors. Nevertheless, there is reasonable agreement in the risk estimates for the two studies.

3.2.7.3 *Entire Population.* Two sets of assumptions were made in projecting the observed mortality to estimates of lifetime risks for a general population. In the first, the same atomic-bomb survivor risk coefficients (additive and multiplicative), derived as a weighted average over all age groups, were used for all age groups in making the projections. In the second, average risk coefficients for the age group 0 to 9 y, and for the age group 10 to 19 y were used for those age groups, while an average "adult" coefficient was used for all the older age groups.

The estimate of the lifetime risk of radiation-induced leukemias is independent of the method of risk projection used, ranging from 9.3 to 10×10^{-2} Gy^{-1} for persons exposed to 1 Gy to the bone marrow. The 10×10^{-2} Gy^{-1} coefficient was obtained using the multiplicative risk projection model and the data for the two younger age groups. It is presumed that the similarity is due to the fact that most of the radiation-induced leukemias have already occurred. On the other hand, the estimates of the lifetime risk of solid cancers is markedly influenced by the choice between age-specific risk coefficient or a constant population or average risk, and between a multiplicative or additive risk projection model. For example, when age specific risk coefficients and the multiplicative risk projection model are applied to the data for those exposed in childhood, the risk estimate is 9.7×10^{-2} Gy^{-1} compared to 6.1×10^{-2} Gy^{-1} when the age-averaged coefficient is used. However, when the additive projection model is used, the choice between the age-specific and age-averaged coefficient makes little difference, 3.2×10^{-2} Gy^{-1} and 3.6×10^{-2} Gy^{-1}, respectively.

The lifetime excess risk of solid cancer for the adult population (≥ 25 y) using the multiplicative risk projection model is 4.7×10^{-2} Gy^{-1}, compared to 9.7×10^{-2} Gy^{-1} if the coefficient for those exposed when children is used. In contrast, with the use of the childhood additive risk coefficients there is no significant difference in lifetime excess mortality between the entire population and the adult only population (3.2 versus 3.6×10^{-2} Gy^{-1}). The differences should signal caution about the method of projecting risks over time in those exposed in childhood. The doubling of the estimated risk for the entire population, compared with the adult population (7.4 versus 4.7×10^{-2} Gy^{-1}) stems from the use of the following age-specific excess relative risk coefficients at 1 Gy of organ dose for all malignancies except leukemia: adults only, 0.35; children 0 to 9, 1.56; and children age 10 to 19, 0.96 (Shimizu et al., 1988; 1990).

The difference between the risk coefficients for adults and for the children 0 to 9 y of age is a factor of 4.5. Evidently the coefficients for the children (0 to 9 y of age) were derived over the entire observation period. However, the estimated excess relative risks for children 0 to 9 y are 1.16 and 1.05 for 1966 to 1975 and 1976 to 1985, respectively, compared with 1.56 for the entire 40 y of follow-up; and for children 10 to 19 y in those periods 0.80 and 0.92 compared with 0.96 for the entire 40 y follow-up (Shimizu et al., 1988). Perhaps it would have been more appropriate to project the childhood relative risks for the future decades as the average for these last two decades, rather than apply the earlier average as a constant excess relative risk over the entire lifetime of these children.

3.2.7.4 *Worker Populations.* The UNSCEAR (1988) lifetime risk estimates for a worker population (Table 3.3) employ the age group 25 to 64 y, whereas BEIR V (NAS/NRC, 1990) used the age group 18 to 65 years. Table 3.3 shows the additive and multiplicative projection models employing age-averaged coefficients for ages 20 to 29, 30 to 39 and ≥ 40 y and age-specific risk coefficients, respectively. For a worker population, using age-specific coefficients, a range of excess cancer risk per 10,000 person-Gy is from 600 (additive model) to 700 (multiplicative model), and using age-averaged coefficients gives a range from 400 (additive model) to 800 (multiplicative model). There is a substantial difference between the estimates for the worker population compared with an "adult population." The difference reflects the inclusion in the latter population of persons irradiated at ages older than age 64 years. Exclusion of these persons means that the working population will contain a larger proportion of persons irradiated in the age groups 25 to 64 than the adult population, and this fact should increase both the additive and multiplicative risk projections for the working population. However, for the additive model there is a decrease and in the multiplicative projection an increase from 700 to 800. The increase in the multiplicative projection is partly due to the increase in the baseline cancer rate for ages greater than 64 years.

The influence of the selection of age-specific risk coefficients instead of age-averaged risk coefficients is shown in Table 3.9 for loss of life span in a working population. Age-specific coefficients are the method of choice particularly for a general population

TABLE 3.9—*Projections of years of life loss for 1,000 persons (500 females and 500 males) exposed to 1 Gy of low-LET radiation at high-dose rate in a working population aged 25 to 64 (adapted from Table 71, Annex F, UNSCEAR, 1988).*

	Risk projection Model	Life lost (y)
Age-specific coefficients	Additive	1,330
	Multiplicative	820
Age-averaged coefficients	Additive	880
	Multiplicative	970

because the childhood risks are significantly higher than in adults (Shimizu *et al.*, 1988).

3.2.8 *Latent Periods*

It is noted in the UNSCEAR Report (UNSCEAR, 1988) that the minimum latent period used for the lifetime-risk projection (*i.e.*, the beginning of the integration period for risk) was 10 y for all cancers except leukemia and bone cancer. The minimum time to tumor diagnosis for an excess risk is among the most difficult quantities to determine if there is a non-negligible baseline risk. The time at which an excess becomes established statistically reflects the minimal latent period, but also the level of excess risk and the rates at which the baseline and the indistinguishable excess cancers accumulate in the study population. Thus, in the atomic-bomb survivors, about 15 y were required before a statistically significant excess mortality was observed for all cancers as a group, but it took 15 to 19 y for stomach cancer, 20 to 24 y for cancer of the lung and breast, 25 to 29 y for cancer of the ovary, and 30 to 34 y for cancers of the colon, bladder and multiple myeloma to appear in excess. The minimum latent period for each of these sites is undoubtedly shorter than the above time periods. However, the assumption of a minimum 10 y latent period for most sites is not unreasonable in the light of available information. The BEIR V (NAS/NRC, 1990) used a minimum latent period of 5 y in their analysis of the thyroid cancer data (see Section 10).

4. Committee on Biological Effects of Ionizing Radiations Reports

Prior to 1990, NAS/NRC produced a number of reports dealing with the effects of ionizing radiation on the health of populations (NAS/NRC, 1972; 1980; 1988).[2] In 1990, the latest of the NAS/NRC series of reports known as BEIR V (NAS/NRC, 1990) was published.

4.1 Earlier Reports of the Committee on Biological Effects of Ionizing Radiations

There have been three major reports from NAS/NRC dealing with the effects of ionizing radiation on the health of populations in a general way. In 1972, the first BEIR committee questioned the establishment of single upper limits for individual and population average exposures as the basis for radiation protection. The BEIR committees in 1972 and 1980 discussed radiation exposures from natural sources, medical procedures and atomic bombs, and evaluated the probability of induction by radiation of genetic damage, cancer and other effects. In 1972, BEIR I (NAS/NRC, 1972) reported an estimate of an annual excess in cancer deaths of about 3,500 in the United States population of as a result of continuous exposure to 1 mSv y^{-1} based on a linear-dose response using an absolute risk-projection model. This estimate was controversial. The report drew from studies of various irradiated populations, but at that time almost all the excess cancer mortality in the Japanese atomic-bomb survivors was from leukemia. Therefore, risk estimates for solid cancers were based mainly on partial-body irradiation.

---

[2]Earlier reports of the NAS/NRC dealing with the biological effects of ionizing radiation were called the Biological Effects of Atomic Radiation Reports, or BEAR Reports, the earliest of which were reports on genetics, pathology, meteorology, oceanography and fisheries, agriculture and food supplies, and disposal and dispersal of radioactive waste which were first published in 1956 and again in 1960.

The BEIR III Report (NAS/NRC, 1980) reflected the difference in opinion about dose-response models that was appropriate for estimating risks of radiation-induced cancer. The report indicated that the committee did not know "whether dose rates of gamma or x rays of about 100 mrads/yr[3] are detrimental to man." At higher dose rates, such as "a few rads per year over a long period," it was considered "that a discernible carcinogenic effect could become manifest." The committee considered the question of the influence of dose rate on the probability of radiation-induced cancer. Based on the linear-quadratic dose-response model, the estimate of excess mortality from all types of cancer for a single whole-body absorbed dose of 0.1 Gy low-LET radiation was in the range of 0.47 to 1.4 percent of the naturally occurring cancer mortality depending on whether the additive or multiplicative risk-projection model was used. Similarly, the estimates of the excess cancer mortality for continuous lifetime exposure to 10 mGy y^{-1} ranged from 2.8 percent to 7.2 percent (Table V-22, BEIR III, NAS/NRC, 1980).

The 1980 report recognized the importance of the heterogeneity of populations with regard to their response to irradiation. The question of susceptible subpopulations and their importance remains today.

Estimates of genetic risk were made. For example, it was concluded that in the first generation, 10 mSv of parental exposure (in the general population) would result in an increase of 5 to 75 additional serious genetic disorders per 10^6 live-born offspring.

The effect of intrauterine irradiation in the development of the brain was discussed and the dependence of the severity of the effect on gestational age was noted. For example, a 28 percent incidence of microcephaly in children irradiated by the atomic bombs in Japan between 4 to 13 weeks gestational age was noted. In this study, a head was considered to be abnormally small when it was two standard deviations below the mean circumference.

Another major BEIR report, BEIR IV (NAS/NRC, 1988) was devoted to the health effects of radon and other alpha-particle emitters and it is discussed in some detail in Section 9 of this Report.

4.2 The 1990 Report of the Committee on Biological Effects of Ionizing Radiations

The BEIR V Report (NAS/NRC, 1990) updates earlier estimates of the risks of somatic and genetic effects of low-level irradiation,

---

[3]100 mrads/yr = 1 mGy y^{-1}.

taking into account new information obtained during the decade since the completion of the BEIR III (NAS/NRC, 1980) study. The report consists of an executive summary, and chapters on background information and scientific principles, genetic effects of radiation, mechanisms of radiation-induced cancer, risk of cancer (including all sites, radiogenic cancer at specific sites), other somatic and fetal effects, and low-dose epidemiological studies. These are basically the same topics treated in BEIR III (NAS/NRC, 1980), the previous report on the effects of exposure to low-LET radiation; however, the newest report differs from BEIR III (NAS/NRC, 1980) in a number of important particulars.

4.2.1   Sources of Data

For estimating the risks of carcinogenic effects, both UNSCEAR (1988) and BEIR V (NAS/NRC, 1990) placed primary reliance on the atomic-bomb survivor data, since they provide the most comprehensive body of information on the effects of a wide range of reasonably well quantified doses of whole-body irradiation in persons of both sexes and all ages. Data are available from a number of other irradiated populations. In particular, for leukemia and cancers of the breast and thyroid (see Table 4.1), comparisons have been made between the risk estimates based on a number of irradiated populations. The occurrence of cancers of the lung, liver and bone, as well as leukemia, as a result of exposure to internally deposited alpha-particle emitting radionuclides, were analyzed separately in BEIR IV (NAS/NRC, 1988) and included in the BEIR V Report (NAS/NRC, 1990).

The atomic-bomb survivor experience was assessed with the use of a machine-readable database, obtained from RERF, comprising numbers of cases and person-years, with average doses for a total of 8,714 cells defined by age at exposure, time after exposure, DS86 dose, city and sex.

4.2.2   Dosimetry

The cohort used by the BEIR V (NAS/NRC, 1990) was the 75,991 Japanese atomic-bomb survivors on whom DS86 dose estimates were available in 1987. Ten categories of absorbed dose, i.e., 0, 0.01 to 0.05, 0.06 to 0.09, 0.1 to 0.19, 0.2 to 0.49, 0.5 to 0.99, 1.0 to 1.99, 2.0 to 2.99, 3.0 to 3.99 and 4+ Gy, were used in the stratification of sex- and city-specific mortality data. The gamma and neutron kerma

TABLE 4.1—*Populations that are the sources of estimates of risks of radiogenic cancers (from Table 4-1, NAS/NRC, 1990).*

Study Population	Source	Incidence or Mortality	Cancer Sites	Total Cases	Total Person Years
Atomic-bomb survivors	Shimizu et al., 1988	Mortality	All	5,936[a]	2,185,335
	Tokunaga et al., 1987	Incidence	Breast	376[b]	940,000
Ankylosing spondylitis patients	Darby et al., 1987	Mortality	Leukemia	36[c]	104,000
			All except leukemia and colon	563[c]	104,000
Canadian fluoroscopy patients	Miller et al., 1989	Mortality	Breast	482	867,541
Massachusetts fluoroscopy patients	Hrubec et al., 1989	Mortality	Breast	74	30,932
New York postpartum mastitis patients	Shore et al., 1986	Incidence	Breast	115	45,000
Israel *tinea capitis* patients	Ron and Modan, 1984	Incidence	Thyroid	55	712,000
Rochester thymus	Shore et al., 1985	Incidence	Thyroid	28	138,000

[a]For both cities, both sexes and all ages ATB for shielded kerma dose in the range of 0.01 to 4.0+ Gy, see Table 4.2 for sites and included types of cancer.
[b]For both cities and all ages ATB, kerma dose in 0.5+ Gy.
[c]Observed cases in males and females treated with x rays.

values in each age, sex and city category were person-years weighted average for the survivors at risk. The DS86 dose to the colon was used for all cancers except leukemia. For leukemia, the dose was based on the DS86 dose to bone marrow. An RBE of 20 for neutrons was used in the calculation of an organ specific dose equivalent for each stratum in the dose-response regressions. Other RBE values were evaluated to assess their effect on the risk estimates. It was considered that the value of the RBE that was used made little difference because of the low doses of neutrons involved.

4.2.3   Data Analysis

The rates of mortality from various forms of cancer were analyzed in relation to age at irradiation, time after irradiation, sex, dose and other variables, with a view toward their compatibility with relative (multiplicative) and absolute (additive) risk models. Parameter estimates for these models were then derived from the data, and the results were evaluated for goodness of fit. To assure sufficient numbers of cases for adequate modeling, cancer deaths were ultimately combined into the following five diagnostic categories: leukemia, excluding chronic lymphocytic leukemia; cancer of the breast; cancer of the digestive system; cancer of the respiratory tract; and other cancers. Total cancers and all cancers other than leukemia were also modeled.

Various exposure-time-response models were fitted to the data, using the Additive Multiplicative Fit (AMFIT) computer program, which fits a general form of the "Poisson regression" model.[4] The observed number of events in each cell of the cross-tabulation was thus treated as a Poisson variate with parameters given by the predicted number of events under the model; i.e., the product of the person-years in that cell times the fitted rate. Although the BEIR III Report (NAS/NRC, 1980) included risk estimates for cancer derived by the constant additive risk-projection model, as well as by the constant multiplicative risk-projection model, the BEIR V (NAS/NRC, 1990) found neither model to fit the data unless modified appropriately for sex, age at irradiation and time after exposure. Given such modifications, the multiplicative risk model was found to provide a more parsimonious description of the data and to be less subject to error resulting from misclassification of causes of

---

[4]This computer program for the analysis of data for cohort survival was developed by D.L. Preston, J.H. Lubin and D.A. Pierce (see EPICURE User's Guide, Seattle Hirosoft Corporation, 1991).

death as discussed below. Hence the BEIR V (NAS/NRC, 1990) preferred risk estimates (Tables 4.2 and 4.3) were based on a modified multiplicative risk-projection model. The committee rejected the additive risk-projection model because it did not fit the data as well.

4.2.3.1 *Comparison of the Models of the 1980 Committee on Biological Effects of Ionizing Radiations and the 1990 Committee on Biological Effects of Ionizing Radiations.* The committee's preferred modeling of lifetime risk of cancer departs substantially from that of the BEIR III Report (NAS/NRC, 1980), and for some cancers involves a modified relative risk projection incorporating time since exposure in addition to the more customary variables, such as sex

TABLE 4.2—*Estimated excess lifetime mortality from cancers of various organ systems after 0.1 Sv acute whole-body exposure, in relation to age at exposure and sex (adapted from Table 4-3, NAS/NRC, 1990).*[a]

Age at Exposure	Males (deaths per 10^5)					
	Total	Leukemia[b]	Nonleukemia[c]	Respiratory	Digestive	Other
5	1,276	111	1,165	17	361	787
15	1,144	109	1,035	54	369	612
25	921	36	885	124	389	372
35	566	62	504	243	28	233
45	600	108	492	353	22	117
55	616	166	450	393	15	42
65	481	191	290	272	11	7
75	258	165	93	90	5	—
85	110	96	14	17	—	—
Average[d]	770	110	660	190	170	300

Age at Exposure	Females (deaths per 10^5)						
	Total	Leukemia[b]	Nonleukemia[c]	Breast	Respiratory	Digestive	Other
5	1,532	75	1,457	129	48	655	625
15	1,566	72	1,494	295	70	653	476
25	1,178	29	1,149	52	125	679	293
35	557	46	511	43	208	73	187
45	541	73	468	20	277	71	100
55	505	117	388	6	273	64	45
65	386	146	240	—	172	52	16
75	227	127	100	—	72	26	3
85	90	73	17	—	15	4	—
Average[d]	810	80	730	70	150	290	220

[a]Based on a single exposure to radiation to a United States population having death rates for 1979 to 1981.

[b]Estimates for leukemia are based on the use of a fitted linear-quadratic model which has an implicit DDREF of approximately two. Estimates for solid tumors are based on the use of a linear model and therefore do not include a DDREF.

[c]Based on the sum of cancers of respiratory tract, digestive tract, breast and other organs.

[d]Averages are weighted for the age distribution in a stationary population having United States mortality rates. Values rounded to nearest ten.

TABLE 4.3—*Projected lifetime excess cancer mortality and associated loss of life expectancy from continuous whole-body irradiation in a population of both sexes (adapted from Table 4-2, NAS/NRC, 1990).[a]*

	Exposure Throughout Life at 1 mSv y^{-1}	Exposure from Age 18 to age 65 at 10 mSv y^{-1}
Lifetime excess cancer deaths		
Number per 10^4	56	300
Percent of normal	3	16
Loss of life expectancy (y)		
Average per person exposed	0.09	0.5
Average per excess death	17	16

[a]Calculations based on cancer and survival rates for the 1980 United States population and on use of the data base presented in Table 4-2, NAS/NRC (1990). Included are an implicit DREF of about two for leukemia and DREF of one for solid tumors.

and age ATB. Thus, it differs from the absolute and relative risk models used in the BEIR III Report (NAS/NRC, 1980) and from the constant relative risk model used by the UNSCEAR (1988) in not assuming that the relative risk is constant with time since exposure. It should be noted, however, that the constant relative risk model was deemed adequate for digestive cancers, and for lung cancer, where the committee's preferred model did not fit significantly better than the constant relative risk model.

4.2.3.2 *Calculation of Excess Risks.* The BEIR V Report (NAS/NRC, 1990) estimated separate lifetime risks for both the exposed and unexposed populations, and the excess risk was defined as the difference between the risk estimates for an exposed and an unexposed population.

For a given radiation-dose equivalent, d, in sievert (Sv), the individual's age-specific cancer risk $\gamma(d)$ was expressed as:

$$\gamma(d) = \gamma_0[1 + f(d)g(\beta)]. \tag{4.1}$$

where γ_0 denotes the age-specific background risk of death due to a specific cancer for an individual at a given age (it also depends upon the individual's sex), $f(d)$ represents a function of the dose d, which is always a linear or linear-quadratic function [*i.e.*, $f(d) = \alpha_1 d$ or $f(d) = \alpha_2 d + \alpha_3 d^2$], and $g(\beta)$ is an excess risk function observed to depend upon a number of parameters; for example, sex, attained age, age at exposure and time after exposure (NAS/NRC, 1990). The age-specific risk could also be modeled as an additive risk:

$$\gamma(d) = \gamma_0 + f(d)g(\beta). \tag{4.2}$$

Both models gave similar results, as expected, since the function $g(\beta)$ was allowed to depend on age, sex and time since exposure. This would not have been the case, however, if $g(\beta)$ depended only on sex and age at exposure (NAS/NRC, 1990).

"The models were fitted using maximum likelihood, $i.e.$, the values of the unknown parameters which maximized the probability of the observed number of cases (the 'likelihood function') were taken as the best estimates, and, where applicable, confidence limits and significance tests are derived from standard large-sample statistical theory," (NAS/NRC, 1990).

It was expected that the form of the background term (γ_o) might vary considerably among populations at risk and would not be of particular interest in terms of radiation risk. Hence the BEIR V chose not to model it, but rather to estimate the baseline rate nonparametrically by allowing for a large number of multiplicative rate parameters, as is often done when fitting hazard models to ungrouped data (NAS/NRC, 1990).

Each model was then described by BEIR V (NAS/NRC, 1990) in terms of point estimates of its various parameters, their respective standard errors and significance tests, and an overall "deviance" for the model as a whole. Because of the extreme sparseness of the data, comparison of deviance to its degrees of freedom was not used as a test of fit of the model; however, since differences in deviance between nested alternative models (pairs of models for which all terms in one model except one were included in the other) have an asymptotic Chi-square distribution, with degrees of freedom equal to the difference in the degrees of freedom among the models being compared. This test was used to assess the improvement in fit as a result of adding terms to the dose-response function and used repeatedly by BEIR V (NAS/NRC, 1990) to minimize potential over-specification of the models. In other words, the model was not made more complex than justified by the data.

Approximate confidence limits on parameter estimates were constructed in the usual way, by adding and subtracting the standard error times 1.65 (for 90 percent confidence) or 1.96 (for 95 percent confidence). However, in cases where the committee had reason to believe that the use of a normal distribution to estimate confidence limits was not valid, it reported likelihood based limits found by searching the likelihood surface iteratively for the parameter values that led to a corresponding increase in the deviance.

4.2.3.3 Models for the Expression of Risk

4.2.3.3.1 *Model for Leukemia.* For leukemia, the BEIR V (NAS/ NRC, 1990) preferred model was a modified relative risk model

(Equation 4.1) with terms for dose, dose squared, age at exposure, time after exposure (minimum latency of 2 y is assumed) and interaction effects. There was a distinct difference between the risks for individuals exposed before age 20 and those exposed later in life. Within these two groups there appeared to be no effect of age at exposure but simply a different time pattern within each group; hence a simple step function with two steps was found to fit both groups rather well (although its biological implications are far from simple).

The parameters for the leukemia model were as follows:

$$f(d) = \alpha_2 \, d + \alpha_3 \, d^2$$

$$g(\beta) = \exp[\beta_1 I(T \le 15) + \beta_2 I(15 < T \le 25)] \text{ if } E \le 20 \quad (4.3)$$

$$g(\beta) = \exp[\beta_3 I(T \le 25) + \beta_4 I(25 < T \le 30)] \text{ if } E > 20,$$

where T is years after exposure, E is age at exposure, and the indicator function $I(T \le 15)$ is defined as 1 if $T \le 15$ and 0 if $T > 15$. The estimated parameter values and their standard errors, in parentheses, are:

$$\alpha_2 = 0.243(0.291), \ \alpha_3 = 0.271(0.314),$$

$$\beta_1 = 4.885(1.349), \ \beta_2 = 2.380(1.311), \quad (4.4)$$

$$\beta_3 = 2.367(1.121), \ \beta_4 = 1.638(1.321).$$

The standard errors for the dose-effect coefficients (α_2, α_3), estimated by means of the likelihood method mentioned above, are both imprecise and the likelihood curves for the estimates are highly skewed.

4.2.3.3.2 *Model for All Cancers Other than Leukemia.* For all cancers other than leukemia and bone cancer, a 10 y latency was assumed; this was done simply by excluding all observations (cases and person-y) less than 10 y after exposure. For purposes of overall nonleukemia-cancer-risk estimation, the BEIR V (NAS/NRC, 1990) simply chose to sum the risks of the components of the nonleukemia cancer group (*i.e.*, respiratory cancer, digestive cancer, etc.), each of which was estimated by the models described below.

4.2.3.3.3 *Model for Cancer of the Respiratory Tract.* For cancer of the respiratory tract, the committee's preferred model was a modified relative risk model (see Equation 4.1) with terms as follows:

$$f(d) = \alpha_1 d \quad (4.5)$$

$$g(\beta) = \exp[\beta_1 \ln(T/20) + \beta_2 I(S)],$$

where T is years after exposure and $I(S)$ is 1 if female, 0 if male, with α_1 is 0.636(0.291), β_1 is $-1.437(0.910)$, β_2 is 0.711(0.610). Under the committee's model, as T varies from 10 to 30, $g(\beta)$ varies from 2.78 to 0.55 meaning that the relative risk decreases by a factor of about five over the period of 10 to 30 y post-exposure. The committee noted that few data are available, as yet, on respiratory cancer among those exposed as children, and that the relative risk is two times higher for adult females (owing to their much lower baseline rates) than for adult males, although the observed absolute excess risks are similar.

When testing departures from a constant relative risk model, the addition of a parameter for time after exposure resulted in the greatest improvement in describing the data for cancer of the respiratory tract, a finding consistent with the decreasing relative risk observed in the ankylosing spondylitis study (Darby et al., 1987), which influenced the committee's choice of parameters. The inclusion of a parameter for *sex* did not improve the model's fit to the data significantly, but caused some improvement. There was no improvement when a term for age at exposure was added to the regression model, its value being sufficiently close to zero as to have no influence on the estimated risk.

4.2.3.3.4 *Model for Cancer of the Female Breast.* For cancer of the female breast, the model was based on a parallel analysis of several cohorts. The important modifying factors were found to be age at exposure and time after exposure. The dependence of risk on age at exposure was complex, doubtless being heavily influenced by the woman's hormonal and reproductive status. Lacking data on these biological variables, the committee found that the best fit was obtained with the use of an indicator variable for age at exposure less than 16, together with additional indicator or trend variables depending on the data set. Both incidence and mortality models were developed. Although these differed, the highest risks were seen in women under 15 to 20 y of age at exposure. Lifetime risk estimates were calculated for mortality only. Risks were very low in women exposed at ages greater than 40, suggesting that risks decrease with age at exposure (see Figure 2.3). Risks were estimated to decrease with time after exposure in all age groups.

The preferred model for breast cancer age-specific mortality (female only) was a relative risk model (Equation 4.1) with terms as follows:

$$f(d) = \alpha_1 d$$

$$g(\beta) = \exp[\beta_1 + \beta_2 \ln(T/20) + \beta_3 \ln^2(T/20)] \text{ if } E < 15 \quad (4.6)$$

$$g(\beta) = \exp[\beta_2 \ln(T/20) + \beta_3 \ln^2(T/20) + \beta_4(E - 15)] \text{ if } E \geq 15,$$

where E is age at exposure and T is years after exposure with

$$\alpha_1 = 1.220(0.610), \beta_1 = 1.385(0.554),$$
$$\beta_2 = -0.104(0.804), \beta_3 = -2.212(1.376), \tag{4.7}$$
$$\beta_4 = -0.0628(0.0321).$$

4.2.3.3.5 *Model for Digestive System Cancers.* For cancer of the digestive system[International Classification of Diseases (ICD) 150 to 159], the most significant aspect of the LSS data was found to be the greatly increased risk (factor of seven) for those exposed under the age of 30, although the committee had no explanation for it. There was no evidence of a significant change in the relative risk with time after exposure.

The committee's preferred model is the relative-risk model given in Equation 4.1 with terms as follows:

$$f(d) = \alpha_1 d$$
$$g(\beta) = \exp[\beta_1 I(S) + \sigma_E] \tag{4.8}$$

where $I(S)$ equals 1 for females and 0 for males and

$$\sigma_E = 0 \text{ if } E \leq 25; = \beta_2(E - 25) \text{ if } 25 < E \leq 35; = 10 \beta_2 \text{ if } E > 35$$

where E is age at exposure. The estimated parameter values are

$$\alpha_1 = 0.809(0.327) \ \beta_1 = 0.553(0.462) \ \beta_2 = -0.198(0.0628).[5]$$

4.2.3.3.6 *Model for Other Cancers.* For cancers other than those listed above, the excess was found to contribute significantly to the total radiation-induced cancer burden. Finer subdivision of the group did not, however, provide sufficient cases for modeling other individual sites. When this was attempted, the models were unstable, resulting in risk estimates for which there was little confidence. The general group of other cancers was reasonably fit by a simple model with only a negative linear effect by age at exposure at ages greater than ten years. There was no evidence of an effect by sex or by time after exposure.

---

[5]There was a typographical error in BEIR V (NAS/NRC, 1990) resulting in the confidence value of 0.327 for the α term being printed as 9.327.

The preferred model is the relative-risk model given in Equation 4.1 with terms as follows:

$$f(d) = \alpha_1 d$$

$$g(\beta) = 1 \text{ if } E \leq 10 \text{ and } \exp[\beta_1(E - 10)] \text{ if } E > 10, \quad (4.9)$$

where E is age at exposure, α_1 is 1.220(0.519) and β_1 is $-0.0464(0.0234)$.

4.2.4 Influence of Dose Rate

Concerning the influence of dose rate on the carcinogenic effectiveness of radiation, the BEIR V Report concluded that low-LET radiation can be expected to decrease in effectiveness when highly protracted, possibly by a factor of two or more for certain neoplasms, if the carcinogenic response of human tissues to exposures at low-dose rate is consistent with that which has been observed in experimental systems (NCRP, 1980; UNSCEAR, 1986). The committee refrained, however, from specifying a precise value for the dose-rate effectiveness factor (DREF), except in the case of leukemia, where its preferred linear-quadratic model (Equation 4.3) contained an implicit DREF of approximately two.

4.2.5 Risk Estimates for Specific Populations

BEIR V (NAS/NRC, 1990) applied the above modified multiplicative risk models to the United States population of 1980. Essentially the multiplicative risk estimates primarily (but not only) from the Japanese atomic-bomb survivors were utilized with the United States baseline cancer statistics by age and sex. Table 4.2 shows estimates of excess lifetime mortality from cancers of various organ systems following an acute dose of 0.1 Sv of low-LET radiation to all body organs for populations of males and of females at specific ages ranging from 5 to 85 years. The committee also estimated lifetime risks of excess cancer mortality for three scenarios of exposure to the United States population with both sexes and two different age distributions. For the general population with all age groups, risks from a single acute 0.1 Sv equivalent dose to all organs and from a continuous lifetime dose of 1 mSv y^{-1} to all organs were estimated. For a working population, with equal numbers of males and females, risks were estimated from a continuous dose of 10 mSv y^{-1} between ages 18 and 65 years. In addition, the loss of life expectancy due to excess cancer mortality was calculated. These results for continuous exposure throughout life are shown in Table 4.3. A comparison of the lifetime excess cancer estimates made by BEIR III (NAS/NRC, 1980) and BEIR V (NAS/NRC, 1990) are given in Table 4.4.

4.2.6 Epidemiological Studies of Populations Exposed to Low Doses

The BEIR V Report (NAS/NRC, 1990) culminates in a brief description of the epidemiological studies, most of them recent, dealing with cancer following exposure to low doses (presumably less than 0.10 Gy) of ionizing radiation. These include individuals residing in areas where the natural background is higher than usually obtains, those receiving occupational exposures (Beral et al., 1988; Gilbert et al., 1989; Kendall et al., 1992; Wing et al., 1991), servicemen exposed in the course of nuclear weapons testing (Caldwell et al., 1983; Darby et al., 1988; 1990; Roman et al., 1987), and individuals residing in the vicinity of nuclear power stations or persons involved in the fabrication of nuclear materials (Darby and Doll, 1987; Jablon et al., 1991; Roman et al., 1987). The BEIR V Report (NAS/NRC, 1990) noted the generally unsatisfactory nature and limitations of these studies and the still limited knowledge on which to base estimates of harm at low doses and low-dose rates. In some of the studies the statistical power is low and in some exposures to other agents

TABLE 4.4—Comparison of lifetime excess cancer-risk estimates from the BEIR III and BEIR V Reports (from NAS/NRC, 1990).

	Continuous lifetime Exposure, 1 mGy y^{-1} (deaths per 100,000)		Instantaneous exposure, 0.1 Gy (deaths per 100,000)	
	Males	Females	Males	Females
Leukemia				
BEIR III	15.9	12.1	27.4	18.6
BEIR V[a]	70	60	110	80
Ratio BEIR V to BEIR III	4.4	5.0	4.0	4.3
Nonleukemia				
BEIR III				
Additive risk model	24.6	42.4	42.1	65.2
Multiplicative risk model	92.9	118.5	192	213
BEIR V[a]	450	540	660	730
Ratio BEIR V to BEIR III	4.8–18.3	4.6–12.7	3.4–15.7	3.4–11.2

[a]These values do not include a reduction factor for low doses whereas the BEIR III values do. The incorporation of a DREF of two in the BEIR V estimates of nonleukemia cancers would reduce appreciably the difference between BEIR III and BEIR V estimates.

as well as radiation confuse the issue of causal relationship between the observed effects and exposure to radiation.

4.2.7 *Other Late Effects*

BEIR V (NAS/NRC, 1990) reported on genetic risks and on some other late effects, which are discussed later in this Report.

4.2.7.1 *Mutagenesis.* This section of the BEIR V Report differed from earlier BEIR reports in three important aspects. First, arguably the biggest departure from previous reports in the estimation of the mutagenic risk of ionizing radiation is a greater reliance on the findings that have emerged from the studies of the offspring of the survivors of the atomic bombing of Hiroshima and Nagasaki. Second, the estimate of the probable lower bound of the DD was raised from 0.5 to 1.0 Sv. Finally, although no specific estimates of risk were made, attention was called to the potential importance of complexly inherited traits.

4.2.7.2 *Other Late and Somatic Effects.* Five areas of concern were addressed under this rubric, namely, (1) cancer in childhood, (2) effects on growth and development following exposure *in utero*, (3) effects of ionizing radiation on cataracts of the lens of the eye, (4) life shortening and (5) fertility and sterility.

5. Comparison of the Reports of the 1988 United Nations Scientific Committee on the Effects of Atomic Radiation and the 1990 Committee on the Biological Effects of Ionizing Radiations

5.1 Data Bases and Analysis

There are many common features in the UNSCEAR (1988) and BEIR V (NAS/NRC, 1990) Reports; a not surprising fact considering that both committees have, by and large, used the same epidemiological data bases and concepts, although they applied different analyses to the data. The cancer mortality of the atomic-bomb survivors was selected by both UNSCEAR (1988) and BEIR V (NAS/NRC, 1990) as the basis for estimating the lifetime risks for the general population and for specific subsets by age at exposure and sex. Both committees contend that the risks of radiation-induced cancer are greater than their respective previous reports estimated (see Table 5.1).

TABLE 5.1—*Estimated excess cancer mortality per 10^4 persons based on a dose of 1 Gy low-LET radiation at high-dose rate to a general population.[a]*

	Risk projection models	
	Additive	Multiplicative
UNSCEAR (1977)	100[b]–250	—
UNSCEAR (1988)	400[c]–500[d]	700[c]–1,100[d]
BEIR III (NAS/NRC, 1980)	170[e]	500[e]
BEIR V (NAS/NRC, 1990)	—	885[e,f]

[a]Japanese population 1982.
[b]Based on risk of leukemia adjusted for low doses and multiplied by five (the assumed ratio of all cancers to leukemia).
[c]Based on age-averaged coefficients.
[d]Based on age-specific coefficients.
[e]Based on linear-dose response for both leukemia and nonleukemia.
[f]Based on United States population 1984.

43

To obtain lifetime risks from data for the first 40 y since the bombing, both committees used risk-projection models. Major questions remain, *e.g.*, (1) how long does the excess risk of leukemias and other cancers last, (2) does the excess risk remain constant or change with time and (3) how is the projection of risk influenced by age at exposure? Both committees agreed that a multiplicative risk-projection is more appropriate than an additive model, at least based on the time course evident so far. The UNSCEAR included the risk estimates based on the additive risk-projection model and noted that the risks estimated by the multiplicative model exceed those obtained with the additive risk-projection risk model. The BEIR V (NAS/NRC, 1990) made no such comparison because it rejected the additive risk-projection model as inappropriate.

The UNSCEAR (1988) chose to use a period of 40 y for the duration of excess mortality from leukemias and a lifetime for solid cancers, using a constant relative risk model. The UNSCEAR Report (UNSCEAR, 1988) based its risk estimates on the data published by Shimizu *et al.* (1987; 1988) and used similar stratification as regards age. The BEIR V (NAS/NRC, 1990) did not use a constant relative risk model alone, but chose a modified multiplicative risk-projection model and stratified for the same variables although somewhat differently than UNSCEAR (1988) in the case of age or time since exposure. The extent of the modification depended on the site and was small for most sites. In the case of the lung, there was significant difference between the estimates of the overall risk using either of the two risk projection models.

The differences in the projected lifetime risk of excess cancer mortality for a population of all ages are not great, 885 per 10^4 persons at 1 Gy by BEIR V (NAS/NRC, 1990) compared to 700 to 1,100 per 10^4 persons at 1 Gy by UNSCEAR (1988) (see Table 5.1).

Excess risk has been determined by the BEIR V (NAS/NRC, 1990) to be the difference between lifetime risk estimates for the exposed and unexposed populations. The United Nations committee (UNSCEAR, 1988), on the other hand, used differences in age-specific rates for the two populations adjusted for the survival of the unexposed population. The approach of the BEIR V (NAS/NRC, 1990) will result in somewhat lower estimates, about 20 percent, of the excess risk than those of the UNSCEAR (1988).

The BEIR V (NAS/NRC, 1990) grouped the RERF data differently than UNSCEAR (1988). The data from the LSS consisted of 8,714 records, reflecting stratification by sex, city, ten exposure groups (based on the kerma at a survivors' location using DS86) and 5 y intervals of attained age, age at exposure and time since exposure.

Most of the analyses by the committee used a reduced data set of 3,399 records collapsing over attained age.

The analyses of the solid cancers were based on the mortality data for persons in the less than 4 Gy group (and who were no older than 75 y of age at death) because there was some evidence of departure from linearity of the dose response above 4 Gy.

To ensure adequate numbers of cancers in relatively specific categories, the cancer deaths were divided into five categories: "leukemia, breast, respiratory, digestive and 'other' cancers." The category of cancers of the digestive system was not more specific because of the inaccuracy of a number of death certificates, especially for cancers of the pancreas and liver, some of which were incorrectly certified as cancer of the stomach.

5.2   Reasons for Increase in Risk Estimates

In 1977, the risk estimates of radiation-induced cancer depended on the estimate of the excess risk of leukemia at low doses and the choice of the ratio of solid cancers to leukemia. As more data for cancer mortality in irradiated populations became available, it was possible to estimate the excess risk of all cancers and of cancer at eight tissue sites as well as leukemia.

Several factors contribute to the increase in risk estimates (see Table 5.1) over those in the 1972 UNSCEAR Report (UNSCEAR, 1972), the 1977 UNSCEAR Report (UNSCEAR, 1977) and the BEIR III Report (NAS/NRC, 1980). One of these is the increased number of cancer deaths in the exposed Japanese population because of the increased period of follow-up, and because of the gradual shift over time of the more susceptible younger cohorts into age ranges at which cancer contributes significantly to overall mortality. When these higher cancer risks are projected over a lifetime the impact on the risk estimates for the general population is substantial. Another contributing factor is the change in the method of analysis, especially the choice of the multiplicative risk-projection model and a linear-dose response for all cancer except leukemia. Also important is the use of DS86 in place of the T65DR (the increase in risk estimates due to the change in the dosimetry varies among tissues). Another element contributing to the increase is the fact that no reduction for low doses or low-dose rates was applied to the risk estimates of solid cancers. Finally, in the case of the United States population, the impact of the use of the multiplicative risk model in the transfer of risk estimates across populations contributes to the increase.

6. Critiques of the Reports of the 1988 United Nations Scientific Committee on the Effects of Atomic Radiation and the 1990 Committee on Biological Effects of Ionizing Radiations

6.1 Data Bases

Although three independent, large epidemiological studies of cancer mortality are analyzed in the UNSCEAR Report, two of them are noted to differ in important aspects:

1. The irradiation was partial body in the case of the ankylosing spondylitis series and the patients with carcinoma of the cervix
2. Both studies were limited with respect to sex (males largely in the ankylosing spondylitis, exclusively females in the carcinoma of the cervix)
3. They were limited to adults (>25 y of age)
4. There was less than lifetime follow-up
5. The generally localized high doses necessitated consideration of cell-killing effects

The comparative relative risk estimates for cancer in various organ sites in the general population of atomic-bomb survivors at 1 Gy (Shimizu *et al.*, 1990) and for secondary primary cancers in women with cervical cancer and treated with radiotherapy (Boice *et al.*, 1985; 1988) are shown in Figures 6.1 and 6.2. In many of the tissues in the cervical cancer patients the organ doses were higher than those incurred by the atomic-bomb survivors. Furthermore,

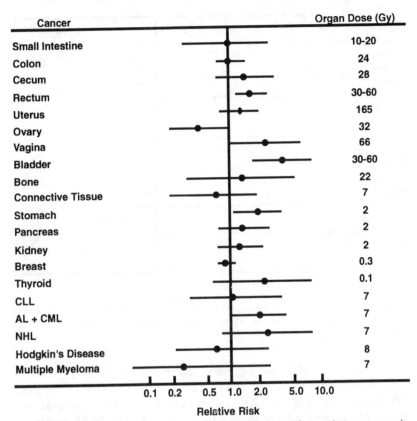

Fig. 6.1. The relative risk of various types of a secondary primary cancer in women treated with radiotherapy for cervical cancer. The 90 percent confidence intervals and the estimated organ dose (Gy) are shown. CLL is chronic lymphocytic leukemia, AL is acute leukemia of all types, CML is chronic myelogenous leukemia and NHL is non-Hodgkins lymphoma (Boice *et al.*, 1985).

both the types of radiation and the time over which the doses were received are different in the two series.

Therefore, the significant conclusions of the UNSCEAR Report are drawn from the most recent follow-up of the Japanese atomic-bomb survivors, employing DS86. However, the risk estimates obtained from the studies of the ankylosing spondylitis and cervical cancer patients were compared and considered consistent with those from the atomic-bomb survivors. The most important data are taken from the RERF LSS Report 11, Part I (Shimizu *et al.*, 1987) and Part II (Shimizu *et al.*, 1988). The risk estimates are, therefore, for the exposed Japanese population and must be transferred to a current Japanese population and then to other populations.

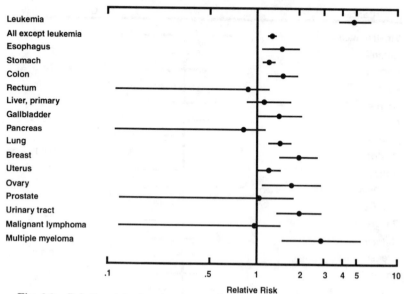

Fig. 6.2.   Relative risk of various types of cancer at 1 Gy shielded kerma and 90 percent confidence intervals for atomic-bomb survivors, 1950 to 1985 (Shimizu *et al.*, 1988).

The incidence of breast cancer is low in the Japanese (Figure 6.3) and their incidence of stomach cancer is high compared to the western population (Figure 6.4). The marked difference in natural incidence of cancer raises problems for projections to other populations using a relative risk model (see later discussion on transfer of risk between populations).

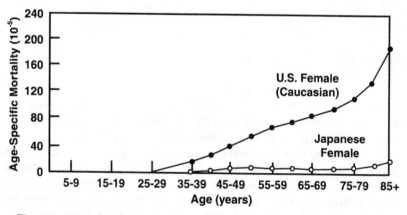

Fig. 6.3.   The age distribution of mortality from cancer of the breast in female caucasians in the United States and females in Japan (from Otake, 1980).

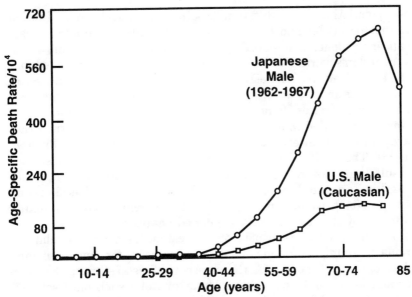

Fig. 6.4. The age distribution of mortality from cancer of the stomach in Japanese male and male caucasians in the United States (Otake, 1980).

The BEIR V Report (NAS/NRC, 1990) implies that several data sets other than those for the atomic-bomb survivors were used in the fitting of the models; however, the bulk of the analysis rests on the atomic-bomb survivor data. Most of the analyses focus on mortality, but where possible, as for example with cancer of the breast, thyroid and skin, incidence data are also considered. There is a need for risk estimates from populations other than the atomic-bomb survivors, especially for cancers of sites, like the stomach, colon and female breast, for which baseline rates are very different between the United States and Japan. With such estimates, it would be possible to evaluate the adequacy of different simple models to transport risk estimates between populations with different baseline rates. For example, in terms of the absolute and relative risk models respectively represented in Equations 4.1 and 4.2, for most ages, the gastric cancer baseline rates, γ_0, are many times greater for a Japanese than for an American population. In applying risk estimates based on the atomic-bomb survivor experience to a United States population, should the estimate of excess relative risk be multiplied by the appropriate United States baseline to estimate excess risk, or should the absolute risk, which is equal to the excess relative risk times the much larger Japanese baseline, be used? Without adequate, and comparable, dose-response data from at least two populations with different baseline rates, such an evaluation cannot be made. The BEIR V (NAS/NRC, 1990) evaluated breast cancer incidence and mortality in the Japanese atomic-bomb survivors and in several North American populations, and concluded that the excess relative risk estimates were in better agreement overall than the estimated absolute risks, and used this finding as the basis for their estimates. It now appears that the incidence comparison was incorrect and that it is the absolute, and not relative, risks that are similar in an evaluation based on more recent data (Preston, 1993). The BEIR V mortality data comparison yielded a similar result that was overridden by the apparently stronger incidence results. Fortunately, the committee's estimates were based mainly on the North American data and were, therefore, only minimally affected by the results of the combined analysis.

The question of transport of risk estimates across populations remains an important one and suitable data sets for comparison and development of models for the transport of risk estimates between populations should be sought. Unfortunately, such data sets are difficult to find for populations and cancer sites for which baseline rates vary widely.

In the BEIR V Report (NAS/NRC, 1990), all of the cancers of the digestive system were analyzed together for the reasons discussed earlier. However, the pooling of the data obscures the important differences between stomach and colon which have very different rates of naturally occurring mortality in different countries. These two sites should be analyzed separately.

In the section on radiogenic cancer at specific sites, the BEIR V Report (NAS/NRC, 1990) considered information from both human and experimental animal studies, which is a good approach and the section provides a very useful compendium of available data. However, the information on at least three sites warrants comment. In the case of skin, the cancer risks are given for incidence, not mortality, and are given separately for areas exposed to sunlight (0.1 excess cases cm^{-2} Gy^{-1}) and those areas shielded from sunlight (0.012 excess cases cm^{-2} Gy^{-1}). In the case of skin cancer in a working lifetime from age 18 to 64 y, the estimated probabilities of incidence and mortality, based on the preferred relative risk projection model, are 9.8×10^{-2} Sv^{-1} and 0.02×10^{-2} Sv^{-1}, respectively (Shore, 1990). The estimate of the risk of fatal skin cancer was obtained by summing the risks for the areas of the skin exposed to ultraviolet radiation from sunlight and shielded areas, averaging risks for both sexes and assuming a lethality of the skin cancers of 0.2 percent (ICRP, 1992). The high risk of radiation-induced skin cancer is of some importance to radiation protection.

In 1980, BEIR III (NAS/NRC, 1980) used a correction of 23 percent on autopsy data to account for under reporting. This correction was not used in the risk estimates derived by the BEIR V (NAS/NRC, 1990) or UNSCEAR (1988) but was used in the projections of risk for total life by Shimizu et al. (1988). In the case of the relative risk model, this correction is not required if the under reporting is uniform for all doses.

6.2 Dose and Relative Biological Effectiveness

In the UNSCEAR Report (UNSCEAR, 1988), the cancer-risk estimates apply to organ doses of 1 Gy of mixed gamma and neutron radiations with a neutron RBE of one. It is estimated in the text that with an RBE of 20, the estimated leukemia risk would be reduced by about 20 percent and the risk of all other malignancies by about 13 percent (Shimizu et al., 1987), but this is not reflected in the tabulated UNSCEAR (1988) risk estimates. Shimizu et al. (1987; 1988) indicate that with an RBE of ten, the risks per Sv will fall by about ten percent for leukemia, and by about seven percent for other cancers. The question of RBE will become increasingly important if the estimate of the contribution of neutrons to the dose at Hiroshima is raised. The report by Shimizu et al. (1987) also notes that if a linear-dose response, up to 4 Gy (instead of 6 Gy), had been used, the risks would increase by 5 percent for leukemia and by 15 percent for other cancers. The reasons for using data only up to 4 Gy appear justified and the corresponding adjustment in risk should be considered.

The BEIR V (NAS/NRC, 1990) defends the selection of an RBE of 20 for neutrons on the basis of an analysis of sensitivity of the effect of varying RBE on the relative risk estimates. The linear coefficient for excess relative risk of leukemia varied from 0.26 per Sv with a RBE of 1 to 0.22 per Sv with a RBE of fifty. For other cancers, the linear coefficient was 1.16 per Sv with a RBE of 1 and 0.97 per Sv with a RBE of 20, or about nine percent less. If, in fact, the RBE was nearer 50, instead of 20, the coefficient would be about 20 percent less.

It is somewhat ironic that the impact of the dosimetry reassessment, which was stimulated partly by the question of neutron RBE, is still dependent on the RBE selected, although only to a minor degree. It should be noted, however, that the RBE value is important mainly for estimates based on the T65DR, which, unlike DS86, had a substantial neutron component of total dose to survivors of the Hiroshima bomb. For example, leukemia risks based on a linear-dose response model and on DS86 bone marrow dose are 2.95, 2.67 and 2.40 excess deaths per 10^4 person-y Sv for RBE of 1, 10 and 20, respectively; whereas the corresponding T65DR estimates are 3.08, 1.81 and 1.23 per 10^4 person-y Sv, indicating that the dependence of the T65DR-based estimates on assumed RBE is far greater than that of the DS86-based estimates. The same is true for breast-cancer mortality; based on the DS86, the mortality estimates are 1.22, 1.00 and 0.82 per 10^4 person-y Sv for RBEs of 1, 10 and 20, whereas the corresponding T65DR estimates are 0.90, 0.43 and 0.26 per 10^4 person-y Sv. However, an RBE of ten has been commonly used for leukemia-risk estimates based upon T65DR, giving a DS86 to T65DR ratio for the leukemia-risk of 2.67/1.81 = 1.48, while for breast cancer a RBE of one was assumed, giving a DS86 to T65DR ratio for the risk of 1.22/0.90 = 1.36. Estimates of risk from the neutron component of the dose should be examined when more cancer mortality data becomes available despite the fact that the estimate will have considerable uncertainty.

6.3 Dose-Response Curves

In general, the use of the DS86 has brought the dose-response curves for cancer mortality for the data from survivors at Hiroshima and Nagasaki closer together. While there is a city difference, it is no longer significant. The dose responses have also become more linear than they were with the T65DR. Both the UNSCEAR (1988) and BEIR V (NAS/NRC, 1990) Reports conclude that the organ-dose response for leukemia is, however, fitted better by a linear-quadratic relationship, modified by a cell-killing term, than by a simple linear relationship. There is a significant reduction in the slope of the response curve for organ doses below 0.5 Gy; specifically, below 0.5 Gy the slope of the excess relative risk is 2.4 Gy^{-1}, while above 0.5 Gy it is 5.5 Gy^{-1}, a change by a factor of about 2.3. The BEIR V Report (NAS/NRC, 1990) provides α and β coefficients for the dose and dose-squared terms in the linear-quadratic relationship, $\alpha D + \beta D^2$, and employs the α term only for low-dose risk estimates. The linear-quadratic relationship also fits better than a linear relationship for colon cancer. For all cancer except leukemia, the best-fitting linear-quadratic equation constrained to have a positive dose-squared term is the boundary solution, i.e., a simple linear relationship. It appears that the linear-dose response was thought to preclude a dose-rate effect. However, there is strong justification from the Japanese data for a reduced leukemia risk at low doses. The excess of cancers in the atomic-bomb survivors at doses below

0.2 Gy does not appear to be significantly different from zero, but can be fitted by either linear or linear-quadratic dose-response curves. However, the data at higher doses suggest that linear-dose-response relationships are more likely to represent the responses for most solid cancers than appeared to be the case in previous reports. The fact that a linear fit appears appropriate for the data over a broad range of doses does not preclude a dose-rate effect. In animal experiments, significant dose-rate effects have been noted for the response of certain tissues in which a linear fit to the data obtained at a high-dose rate was considered the best fit (Ullrich et al., 1987).

6.4 Risks at Low-Dose Rates and with Protracted Exposures

The BEIR V (NAS/NRC 1990) Committee estimated the risks of continuous lifetime exposure to 1 mSv y^{-1} and 0.01 Sv y^{-1} for age 18 to 65 y of age. For a given cancer type, the fitted model, which depends upon age, age at exposure and sex, was used to calculate the relative risk for each age. This estimate was obtained by integrating the relative risk terms up to the given age. Life-table methods were used and the relative risk was multiplied by the base rate and thus the number of expected deaths were obtained and the excess deaths by substraction. This was done for each cancer site including "all other cancers" so that the number of the total excess cancers was obtained. No allowance for a reduced effect on the induction of solid cancers at low-dose rates (e.g., 10 mGy y^{-1}) was made in the risk estimates presented in either the UNSCEAR (1988) or BEIR V (NAS/NRC, 1990) Reports. Both committees maintained that the carcinogenic effect of low-LET radiation is generally less at low doses and low-dose rates compared with those at high doses and high-dose rates, but noted that relatively few human data exist on which a dose-rate-reduction factor could be based. However, as noted above, there is a significant difference in the excess relative risk of leukemia for those exposed to ≥0.5 Gy compared to those exposed to lower doses. This reduced coefficient was not used in any of the UNSCEAR (1988) projected risk estimates. The alternative of employing the initial slope (α) of a linear-quadratic fit to the leukemia data for low-dose assessment was also omitted in the UNSCEAR Report (UNSCEAR, 1988), but was adopted for low doses in the BEIR V Report (NAS/NRC, 1990).[6]

[6]The formula given on page 56 of the BEIR V Report (NAS/NRC, 1990) to describe the calculation of risk resulting from protracted exposures was changed in the second printing of the report. However, neither formula appears to be consistent with the text.

Both the UNSCEAR (1988) and BEIR V (NAS/NRC, 1990) Reports noted that a downward adjustment of the risk by a factor between two and ten, or two or more, respectively, may be indicated. BEIR V (NAS/NRC, 1990) favored a value of two, a figure that is consistent with the estimate of the reduction factor in the leukemia risk at low doses for the atomic-bomb survivors. The UNSCEAR (1988) chose not to select a DDREF and BEIR V (NAS/NRC, 1990) chose not to use one for solid cancers.

6.5 Dose-Rate Effectiveness Factors

Both general and working populations are exposed constantly to a certain level of natural background at a low-dose rate to which, intermittently, are added small increased levels of radiation at a higher-dose rate or protracted low-dose-rate irradiation. Neither committee has given precise guidance of how to deal with these common patterns of exposure, but they do agree that some correction should be made to estimates obtained at high doses and high-dose rates for their application to conditions of protracted exposures.

Since the 1977 UNSCEAR Report, it has been common practice to apply a DDREF for low doses and/or low-dose rates. The DDREF of 2.5 used in the 1977 UNSCEAR Report was adopted by ICRP (1977) and NCRP (1987) and indirectly by BEIR III (NAS/NRC, 1980) by basing the risk on the α coefficient of a linear-quadratic dose response for leukemia. In the NIH Radioepidemiological Tables, a reduction factor of about 2.3 was used based on the nature of the linear-quadratic dose-response relationship (NIH, 1985). In the U.S. Nuclear Regulatory Commission's Report (NRC, 1989), a DREF of 3.3 was applied to estimate the risk at low doses for cancers at all sites except thyroid and breast. In a subsequent NRC Report (NRC, 1991), a DREF of 2.0 was used. Similarly, the National Radiation Protection Board in the United Kingdom considered a DREF of three (NRPB, 1988) for all cancer sites except breast, but have recently decided to apply a DREF of two (NRPB, 1993).

There is support in the UNSCEAR Reports (UNSCEAR, 1977; 1988) for a reduction in leukemia risk at low doses. For example, in UNSCEAR (1988), based on the change in slope of the dose-response curve for the atomic-bomb survivors below 0.50 Gy, a reduction factor of about 2.3 occurs. Recent observations of the induction of thyroid cancer by radiation indicate that at doses of about 0.5 Gy, radiation from ^{131}I is about four times less effective than acute irradiation with x rays, suggesting a DREF of about four (Holm et al., 1988). It has been suggested that spatial distribution of the dose and hormonal factors influence the difference in the effects of protracted irradiation from ^{131}I compared to those of the brief exposure to x rays. The apparent reduction of breast-cancer risk in a Canadian study, following multiple doses of less than 10 mGy compared to multiple doses greater than 10 mGy (Miller et al., 1989), is also noted. Earlier studies had not found any effects of fractionation in these two tissues (Beral et al., 1988; Cardis and Kaldor, 1989; Gilbert et al., 1989; Howe et al., 1987; Kendall et al., 1992; Smith and Douglas, 1986).

There are a number of current studies of radiation workers who have been exposed to radiation over a long period of time. However, there are difficulties in such studies because of the complex exposures and in attaining a level of statistical power necessary to obtain unequivocal evidence of the influence of dose rate and protraction. The lack of data for humans exposed to protracted low-dose-rate low-LET radiation has forced a dependence on the use of data from experimental studies.

NCRP Report No. 64 (NCRP, 1984) examined the effect of dose rate on a wide range of experimental end points, from mutations to life shortening, in experimental animals. Dose-rate effectiveness factors were calculated from the ratios of the coefficient of the linear responses obtained after exposure at high- and low-dose rates. A reduction in effect with reduction in dose rate appeared to hold for all the biological end points examined and the reduction factors ranged between two and ten.

The 1986 UNSCEAR supported a DDREF based on a variety of experimental studies, including cancer induction and life-span shortening in animals exposed to low-LET radiation. The report concluded that linear extrapolation from high-dose data to low doses (less than 0.2 Gy), could result in overestimation of risk by a factor of up to five.

The BEIR V Report (NAS/NRC, 1990) also considered the experimental data and found DREFs that ranged between two and ten for specific locus mutation, reciprocal translocation, tumorigenesis and life shortening in experimental animal studies. For these end points, the committee considered the single best estimates of the DREF were four to five, but they did not use a DREF in their estimates of risk of any solid cancer. However, the report did include a formulation for the estimation of risk of cancer in persons exposed over protracted periods. For example, continuous lifetime exposure to 1 mSv y^{-1} and continuous exposure to 0.01 Sv y^{-1} from 18 until 65 y of age.

The apparent linear-dose-response relationships for solid cancers noted in the UNSCEAR (1988) and BEIR V (NAS/NRC, 1990)

Reports present an interesting problem in the interpretation of what the relationship of the high-dose data is to the estimates at low doses and low-dose rates. It should be noted again that although the dose-response curves for solid cancers are best fitted by a linear model, this does not eliminate the possibility of a dose-rate effect.

Linear responses have been fitted to data obtained over a range of 1 to 4 Gy. In this dose range, there are very many tracks traversing each nucleus. The number of tracks traversing the relevant targets is dependent on dose and the number and size of the targets for induction of cancer, which are unknown. It seems unlikely that at doses of 3 or 4 Gy the relevant targets are traversed by a single track and, therefore, the effects would be dose-rate dependent.

In the past, it has been assumed that the data for the induction of solid cancers, as well as leukemia, could be fitted as given in Equation 6.1:

$$E = \alpha D + \beta D^2 \tag{6.1}$$

where E is the effect, D the dose and α and β are coefficients. When the linear-quadratic model applies, the DREF can be estimated. DREF $= 1 + (\beta/\alpha)D$ (see Figure 6.5). In the case of human leukemia, the estimates of DREF based on the linear-quadratic model are about two to three. BEIR V (NAS/NRC, 1990) considered 2.1 to be the

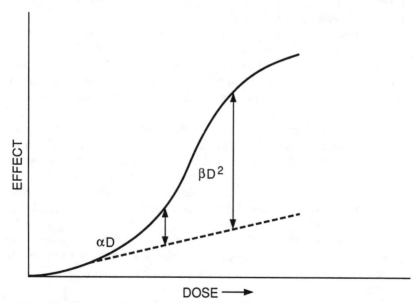

Fig. 6.5. Schematic of effect as a function of radiation dose illustrating the effect of lowering the dose and also dose rate.

single best estimate of the DREF for leukemia. The ICRP (1991) recommended that a DDREF should be "included in the probability coefficients for all equivalent doses resulting from absorbed doses below 0.2 Gy and from higher absorbed doses when the dose rate is less than 0.1 Gy per hour." When exposure to radiation is not only at a low-dose rate, but is protracted over a large fraction of the life span, the change in susceptibility for cancer induction with age becomes important and can contribute to the reduction of effect.

There is a need for more extensive data from humans exposed to multiple fractions of low doses and exposures at low-dose rates. Similarly, experimental animal data will be needed, especially to elucidate the role of dose rate and fractionation in determining the dose-response relationships of the complex process of multistage carcinogenesis. It is clear that the linear-quadratic dose-response model that is appropriate for induction of chromosome aberrations and leukemia may be inappropriate for the task of describing the dose-response relationships of the complex process of radiation-induced solid cancers.

6.6 Age and Sex Dependence

Both committees noted sex-dependent differences in total cancer risk, which were less in the BEIR V (NAS/NRC, 1990) estimates (only five percent) than in the UNSCEAR Report (UNSCEAR, 1988). The sex-dependent differences were considered to be 50 percent by ICRP (1977) and about 40 percent by BEIR III (NAS/NRC, 1980). In the new risk estimates, the sex dependence ranges from about five percent (NAS/NRC, 1990) to about 20 to 40 percent (UNSCEAR, 1988). The sex-dependent differences are age dependent with the young having the greatest dependence. Similarly, there was qualitative agreement about the influence of age at exposure. In the case of the BEIR V Report (NAS/NRC, 1990), the total excess cancer mortality in males decreased from about 1.28×10^{-2} persons Sv^{-1} as a result of an equivalent dose of 0.1 Sv at 5 y of age to about 0.48×10^{-2} persons Sv^{-1} for those exposed at 65 y of age. UNSCEAR (1988) did not stratify for ages greater than 50 years. Estimates of excess cancer deaths by UNSCEAR were about 1.82×10^{-2} persons Gy^{-1} exposed between 0 to 10 y of age and about 0.48×10^{-2} persons Gy^{-1} at ages of 50 or greater at time of exposure.

6.7 Risk-Projection Models

To estimate the lifetime cancer-mortality risks in the atomic-bomb survivors who were young, less than about 45 y of age ATB, it is necessary to project risks into the future. Since the 1970s, there has been increasing support for the constant relative or multiplicative risk-projection model and less support for the constant absolute or additive risk-projection model. This is particularly true for cancer of the lung, breast, stomach and for all cancer except leukemia (see Table 3.2). The major exception to the multiplicative risk-projection model is provided by the rapid decrease in relative risk for leukemia for the two youngest age groups (0 to 9 y and 10 to 19 y ATB) over the first two or three decades following exposure. The recent review of the data by Shimizu et al. (1988) for the period 1950 to 1985 shows a decreasing trend of relative risk after 1965 for the age group 20 to 29 ATB, but not for age groups 30 to 39 y and 40 to 49 years. It may be that neither the constant additive or constant relative risk model is completely correct for projection of risk with time. However, with regard to organ systems (all ages, both sexes), there is an approximate constancy in relative risk with time for all cancers except leukemia. The BEIR V Report (NAS/NRC, 1990) has adopted a generalized multiplicative model in which provision is made for a decline in relative risk beyond a specified time after exposure, in particular for breast cancer and lung cancer. The actual degree of the decline is determined by fitting the model to the data for each tissue.

The UNSCEAR Report (UNSCEAR, 1988) reviewed time projection models (Annex F, in the UNSCEAR Report, paragraphs 112 to 116), and, in particular, the mathematical risk model of Muirhead and Darby (1987). In general, the best-fitting model was intermediate between the multiplicative and additive models, depending on which of the three variables, sex, age at exposure and time after exposure, were included in the model. When age at exposure alone was included, the multiplicative model provided a better fit than the additive model; when age at exposure and time after exposure, or when all three variables were included, the additive model provided the better fit (Table 12, Annex F). Using DS86, Shimizu et al. (1988) have repeated the Muirhead and Darby (1987) investigation and found that the simple constant absolute-projection risk model in which the excess risk depended on age at exposure and sex, but not on attained age after exposure, can be rejected (at a high level of significance), whereas the constant relative risk model fits the data within reasonable limits.

The uncertainty regarding the choice of a model has been recently compounded by the results of the further follow-up of the ankylosing spondylitis series (Darby et al., 1987). In this study it was concluded that the excess relative risk of all neoplasms, other than leukemia or colon cancer, reached a maximum between 10 and 12.4 y after irradiation and the excess declined after about 25 years. This finding suggests that the multiplicative model will overestimate the lifetime risk. As noted earlier there is some evidence for a decline of relative risk for some organs and tissues in the Japanese data, such as lung, esophagus and multiple myeloma, but not sufficiently to cause a decline in the relative risk for all solid cancers. Furthermore, the data from other populations such as the patients examined by fluoroscopy in the Canadian study (Miller *et al.*, 1989), the uranium miners (NAS/NRC, 1988) and the ankylosing spondylitis patients discussed above, suggest that the constant relative risk model used by UNSCEAR (1988) overestimates the risk for at least some solid cancers, for example, breast, stomach and lung. The constant relative risk-projection model does, however, fit most of the data better than the constant absolute risk-projection model.

The risk of leukemia is elevated beyond 25 y after exposure, but at a level well below the peak-mortality rates in both the atomic-bomb survivors and the ankylosing spondylitis patients. However, it is noted that the relative risk of leukemia among the Japanese atomic-bomb survivors has remained elevated for the period 1982 to 1985 at approximately 1.5 and it is possible that the plateau (expression time) may exceed 40 years. If this is the case, the lifetime-risk projection for leukemia using the 40 y plateau would underestimate the actual lifetime risk by about two to three percent, assuming a five percent y^{-1} decrement in relative leukemia risk. Neither the risk estimates for leukemia nor the modeling in these reports reflects the behavior of individual types of leukemia.

The risk-projection models used in BEIR V (NAS/NRC, 1990) in its projection of risk are generally well described and will influence the choice of models in future assessments. However, the report makes certain assertions regarding the correct modeling of lifetime-cancer risks, but fails to distinguish between those findings that are confirmatory, and those that are new. The richer model, that the BEIR V (NAS/NRC, 1990) used, may improve the precision of the risk projections through a better characterization of the observed effects, because it incorporates not only information from the atomic-bomb survivors, but also from other studies, particularly about the change in risk with time after exposure.

It is moot, however, whether the assumptions implicit in this finer modeling are met by the data because although the recent LSS report and the data used by BEIR V (NAS/NRC, 1990) suggest that all deaths through 1985 have been ascertained, this is, in fact, not the case. The *koseki* (obligatory household registries) checks, on which the ascertainment of death ultimately rest, do not all occur in 1 y; they are actually spread over five years. Thus, an individual whose *koseki* was examined in 1981 at the beginning of a 5 y period could have died in 1982, 1983, 1984 or 1985, and this fact would not be learned until his or her *koseki* was perused again in 1986, the next cycle of surveillance. As examination of the number of deaths in a given period of time in successive LSS reports reveal, the mechanics of the surveillance results in substantial late reporting of death. Moreover, it can be argued, and the data support this contention, that since relatively more of the younger individuals resided outside of Hiroshima than Nagasaki, that late reporting occurs and is not independent of the age of the individual ATB. For example, Beebe *et al.* (1978b) describe 736 deaths (both sexes, all ages) from malignant neoplasms in the years 1971 to 1974, but on the next summary of the data 4 y later, Kato and Schull (1982) reported 794 for this same interval. For individuals under the age of 20 y ATB, the deaths numbered 68 and 75, respectively, or an increase of more than ten percent in this time interval, which is important given the small numbers involved (12 versus 15 deaths in the case of individuals 0 to 9 y of age ATB).

Thus, within the LSS the shape of the curve of deaths as a function of time since exposure and age at exposure is obviously influenced by the nature of the follow-up process. It should be noted, however, that this applies more to the distribution of time since exposure for cancers other than leukemia, than to the effort to model time since exposure for leukemia. It is generally presumed that most of the leukemia cases have already occurred and, thus, the distribution of time since exposure is less likely to be biased by late reporting.

Despite the ostensibly better modeling, the BEIR V Report's (NAS/NRC, 1990) lifetime-risk projections do not differ substantially, in terms of excess cases in a lifetime, from the risk projections of the UNSCEAR (1988) Report or those of the NRPB (1988). The BEIR V Report's (NAS/NRC, 1990) risk projections do differ, however, in the pattern of time over which excess cases manifest themselves, but, as implied above, it is arguable how real these patterns are. Nevertheless, their modeling has provided a useful approach that may become more important, with time, if future data indicate declining risks with age from all sites and that the constant relative risk model is clearly inappropriate.

6.8 Transfer of Risks to Other Populations

The problem of transfer of risks to other populations is also discussed in a later Section of this Report in relation to weighting factors. Because of the continuously changing risk factors, and therefore the baseline mortality, it is difficult to give confidence limits on lifetime-risk projections and the problems are compounded with the transfer of risks from one population to another. Estimates of cancer risk associated with exposure to radiation in a specific population have to be based on the experience of other populations that have been exposed. The exposed populations may differ in terms of ethnicity, occupation, health and other relevant characteristics from the population for which risk estimates are required.

Such is the case with risk estimates for the United States population, because the data used by UNSCEAR (1988) and BEIR III and V (NAS/NRC, 1980; 1990) come almost entirely from the Japanese that were in Hiroshima and Nagasaki ATB in 1945. The atomic-bomb survivors belong to the Japanese population of 1945 that differs considerably from both the Japanese and the United States populations of the 1990s and baseline mortality is changing continuously in each population. The UNSCEAR (1988) chose three populations (Japan, United Kingdom, Puerto Rico) to which it applied the absolute and relative risk coefficients obtained from the data for the atomic-bomb survivors exposed to 1 Gy of low-LET radiation at a high-dose rate to estimate the lifetime excess cancer mortality in the three reference populations (see Table 6.1).

The projection of lifetime excess mortality of the three populations for leukemia and other cancers did not differ markedly. This demonstrates that, despite marked differences in rates at specific cancer sites, the overall risk of cancer is surprisingly similar in different populations. When individual cancer sites are considered, the importance of the differences in baseline-mortality rates becomes evident. For all sites, and for specific sites, it can make a very significant difference whether the risk estimates are expressed in terms of absolute or relative risk.

There is not an adequate understanding about the underlying factors involved in the transfer of risk projections across populations

TABLE 6.1—*Comparison of projections of lifetime excess cancer mortality in three reference countries for 10,000 persons of the general population exposed to 1 Gy of organ absorbed dose of low-LET radiation at high-dose rate (adapted from Table 65, Annex F, UNSCEAR, 1988).*[a]

	Japan		United Kingdom		Puerto Rico	
	Multiplicative	Additive	Multiplicative	Additive	Multiplicative	Additive
Leukemia	97	93	130	85	94	97
Other cancers	610	360	630	310	490	400
Total	707	453	760	395	584	497

[a]Excess risk coefficients derived from atomic-bomb survivors.

and, therefore, it is difficult to make a categorical recommendation on what method of transfer of risks should be used.

6.9 Estimates of Risk for Working Populations

The UNSCEAR (1988) estimates of lifetime risk to working populations are unsatisfactory for assessment by NCRP of the risk for chronic occupational exposures for several reasons:

1. The lower age level for this population should be 18 rather than 25, (this will increase the risk for two reasons; a possible increased sensitivity for the 18 to 25 y of age cohort, and there will be a greater number of years at risk after exposure at these ages)
2. No specific dose or dose-rate effectiveness factor has been proposed
3. All lifetime-risk estimates are for a single dose at a defined age. Therefore, it appears that the estimated lifetime risks (shown in Table 3.3) based on risk estimates for populations exposed acutely to 1 Gy have uncertain relevance to the populations of concern for radiation protection

There are no estimates of risk in the UNSCEAR (1988) Report that are applicable to occupational (fractionated) exposure at low doses and low-dose rates, either in the form of coefficients per unit equivalent dose (such as 0.01 Sv) or lifetime risk from repeated annual exposure. Lifetime risks, following the examples in BEIR III (NAS/NRC, 1980) using both risk-projection models (see Section 6.11 and Sinclair, 1985) could be made. The use of the multiplicative risk-projection model involves a dependency on time since exposure, where the annual increment of cancer risk for a given dose increment will increase as a function of time, and reach a maximum before declining for certain types of cancer, as discussed in BEIR V (NAS/NRC, 1990). On the other hand, the age at exposure dependency of relative risk falls as a function of age, which will also modify the lifetime risk. An example is given in BEIR V (NAS/NRC, 1990) for a working population receiving 0.01 Sv per y from the age of 18 to the age of 65, but no dose-rate effectiveness factor is applied for cancers other than leukemia.

6.10 Competing Causes of Mortality and Independence of Risks

For previous projections of lifetime-cancer risk, it has been assumed that the only radiation-related risk of mortality was that attributable to cancer. Alternatively put, it has been assumed that competing causes of death were independent of that associated with the increased frequency of cancer following exposure. Until recently, this assumption seemed defensible on the basis of the studies of the atomic-bomb survivors, but now data are emerging that suggest it may no longer be tenable at high doses. If the competing causes of death are not independent, past estimates of lifetime risk are in error, possibly substantially so. Yashin *et al.* (1986) have demonstrated the importance of accounting for dependence among causes of death in assessing the impact of eliminating one cause of death on overall mortality rates. Neither the UNSCEAR (1988) nor the BEIR V (NAS/NRC, 1990) Reports considered the problem of independence. The problem of association of cancers, especially in studies of irradiated female mice, has long been recognized (Storer, 1982). The statistical issues involved are complex (see Cox, 1962; Heckman and Honore, 1989; Schatzkin and Slud, 1989; Tsiatis, 1975) and suitable models often presuppose information that does not presently exist, for example, the form of the dependency. While it is unlikely that answers to this problem will be forthcoming quickly, it would seem appropriate to begin their exploration, and if need be, enter the appropriate caveats after the projections are made.

6.11 Summary of New Risk Estimates for Cancer Mortality

The estimates of the risk of cancer mortality as a result of radiation exposure have increased compared to those made by UNSCEAR (1977) and by BEIR III (NAS/NRC, 1980). The UNSCEAR (1988) estimated lifetime risks range from 4×10^{-2} Sv^{-1} excess fatal cancers for a Japanese population of all ages and both sexes based on age-specific coefficients using an additive risk-projection model to 11×10^{-2} Sv^{-1} based on age-specific coefficients and a multiplicative risk-projection model. These estimates have made no allowance for a decrease in the effect for low-dose-rate exposures. Depending on whether a DREF of two or three is used, the range of risk estimates for low doses or low-dose-rate irradiation would be 1.3×10^{-2} Sv^{-1} to 5.5×10^{-2} Sv^{-1}. The relatively broad range reflects the number of factors that influence the risk estimates, for example, age distribution, dose rate and the method of analysis. The BEIR V (NAS/NRC, 1990) risk estimate of 8.9×10^{-2} Sv^{-1} for high-dose-rate exposures of a general population is based on a modified multiplicative model with a linear-quadratic dose response for leukemia and a linear-dose response for cancers other than leukemia. The risks were estimated from the atomic-bomb survivors but projected to the United States population.

BEIR V (NAS/NRC, 1990), in contrast to UNSCEAR (1988), made a risk estimate of 6.3 × 10^{-2} Sv^{-1} that might be incurred by a working population (18 to 65 y of age) with a protracted exposure (0.01 Sv y^{-1}). This includes an implicit DREF of about two for leukemia, derived from the linear-quadratic dose response of 0.01 Sv. If a DREF of two were also applied for all nonleukemia cancers, the total risk would be about 3.5 × 10^{-2} Sv^{-1} and 3 × 10^{-2} if a DREF of 2.5 were applied. This estimate for the worker population risk has the advantage that it is based on an appropriate age range and is for the United States population.

The application of a DREF of two for solid cancers is consistent with the recommendations of both UNSCEAR (1988) and BEIR V (NAS/NRC, 1990). The low-dose-rate estimates of 3.5 × 10^{-2} Sv^{-1} and 3.0 × 10^{-2} Sv^{-1} can be compared with that of 1.25 × 10^{-2} Sv^{-1}, the average of the risks estimated for the two sexes in 1977 (ICRP, 1977). It is clear that all the recent estimates exceed those previously made. The reasons for the increase in risk estimates for the atomic-bomb survivors have been discussed earlier (see Section 4.2).

In general, the increase in risk estimates appears to be soundly based. However, the differences in the estimates of leukemia mortality reported by BEIR III (NAS/NRC, 1980) and BEIR V (NAS/NRC, 1990) are striking (see Table 4.4) and require examination.

The BEIR V (NAS/NRC, 1990) leukemia estimates are about four times greater than those of BEIR III (NAS/NRC, 1980) for both females and males. The increase in the estimate by UNSCEAR is about a factor of two. The differences are surprising, as the number of cases of leukemia that have appeared in the atomic-bomb survivors since 1976 is small. The reasons for the differences in the estimated risks include the new DS86 and the neutron RBE used. Based on DS86 and an RBE of 20 for neutrons that was used by the BEIR V (NAS/NRC, 1990), the estimate of excess deaths for leukemia per 10^{5} persons per 0.1 Sv is 2.2 times the previous estimate based on the T65DR. When the data are restricted to doses below 4 Gy (see Figure 6.6 for the number of cases of leukemia at the various doses), and the RBE of 27.8 applied to the low doses in BEIR III (NAS/NRC, 1980) is taken into account, the differences between T65DR and DS86 estimates become closer to three. In the BEIR III Report (NAS/NRC, 1980), the risk at low doses was estimated on the basis of the linear-quadratic model using the so-called "cross-over dose." This is the dose at which the contribution of the linear and quadratic coefficients are equal and was estimated to be 1.16 Gy whereas the estimate based on the BEIR V (NAS/NRC, 1990) analysis is 0.89 Gy.

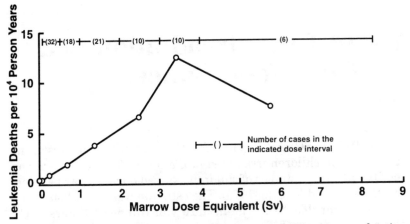

Fig. 6.6. Cumulative leukemia mortality in Nagasaki and Hiroshima as a function of the estimated equivalent dose in Sv to the bone marrow using a quality factor of 20 for the neutron component, BEIR V (NAS/NRC, 1990).

Similarly, the estimate of the DREF, based on the linear-quadratic dose-response relationship for leukemia in the BEIR III Report (NAS/NRC, 1980) was about 25 percent greater than that of the BEIR V Report (NAS/NRC, 1990). Other sources of the differences between the estimates made in 1980 and 1990 are: (1) The inclusion of chronic lymphocytic leukemia in the United States population spontaneous rate used in the calculation by the BEIR V (NAS/NRC, 1990) estimate, but not in the BEIR III (NAS/NRC, 1980) estimate, and in the differences in the models used for analyses; and (2) BEIR V (NAS/NRC, 1990) made an estimate of the leukemia mortality for the years of 1945 to 1950 which it was suggested may be an underestimate, but, in fact, it may be an overestimate.

7. *In Utero* Irradiation and Childhood Cancer

The 1977 UNSCEAR Report (UNSCEAR, 1977) estimates of the cancer risk in children irradiated *in utero* were based primarily on the Oxford survey studies of children exposed during x-ray pelvimetry (Stewart and Kneale, 1970). The estimate was about 200 to 250 excess cancer deaths in the first 10 y of life per 10^4 person-Gy absorbed fetal dose. Fifty percent of these malignancies were childhood leukemias and 25 percent were tumors of the central nervous system. In the most recent 1950 to 1985 follow-up of the Japanese atomic-bomb survivors, no childhood leukemias were identified. Two cases of cancer, one of the liver and the other a Wilm's tumor, occurred in heavily exposed individuals during the first 14 y of life (Yoshimoto *et al.*, 1988). A total of 13 cancers, including two leukemias, mainly in adults, were found in the 0.01 to 2.2 Gy dose groups up to 39 years of age. This results in a risk of about 230 excess cancer cases per 10^4 person-Gy, which is comparable with a risk estimated from a reanalysis of the Oxford survey data of approximately 220 excess cancer deaths per 10^4 person-Gy (Bithell and Stiller, 1988), although the cancer spectrum and age range of the cases are very different.

The average age at the onset of cancer in the atomic-bomb survivors that were exposed *in utero* was about 25 y of age. There were five cases of cancer in the zero-dose groups, two of the uterus and three of the breast, which appeared at an average age of about 33 years. The number of cases is small and the types of cancer are different between the exposed and control groups. In contrast to the control group, there were no cases of breast or uterine cancer in the irradiated groups. There is a question about the causal relationship between radiation and some of the types of cancer and, of course, the risk estimate should be based only on radiogenic cancer.

Mole (1974) observed in England that the relative risks of childhood leukemia and other cancers in irradiated twins, compared with nonirradiated twins, were elevated; these findings were supported by a similar study in Connecticut (Harvey *et al.*, 1985) and one in Sweden (Rodvall *et al.*, 1990). A large cohort study was carried out by Monson and MacMahon (1984) and the authors concluded that there was an excess risk only for leukemia and only for the first 8 y of life. Although these epidemiological studies suggest that an association exists between *in utero* exposure to diagnostic x rays and childhood cancers (Mole, 1974), both UNSCEAR (1988) and BEIR V (NAS/NRC, 1990) concluded that precise estimates are difficult to calculate, and the magnitude of the risk remains uncertain.

The UNSCEAR, in two reports (UNSCEAR, 1986; 1988), reviewed the cancer risks of the irradiated embryo/fetus and discussed the strengths and limitations of the scientific literature. Epidemiological studies have shown an association of childhood cancer with prior irradiation, but have left doubts as to whether radiation is causally involved, in view of the confounding factors. While there are problems with all the epidemiological studies, they suggest a relative risk of cancer of about 1.4 or 1.5 following prenatal diagnostic x-ray exposure, with no reliable data on dose-response relationships. The UNSCEAR Report (UNSCEAR, 1986) concluded, based on the epidemiological surveys of prenatal medical x-rays, the Japanese atomic-bomb experience, animal experiments and the cancer biology involved, that existing data cannot resolve the question of the quantitative relationship between prenatal irradiation and childhood cancer. In addition, it was noted that the causal relationship between prenatal irradiation and leukemogenesis had not been established beyond doubt. Accordingly, the UNSCEAR (1988) Report concluded that the studies on childhood induction of cancer following prenatal exposure to radiation provide evidence that there are effects, but the data do not provide risk coefficients useful in risk projection for childhood cancers.

8. Radiation Effects in Children of Irradiated Parents

The risk of excess cancer up to the age of 20 y has been studied in the children born to atomic-bomb survivors (Yoshimoto *et al.*, 1990). Gonadal doses were calculated using DS86 and assuming an RBE of 20 for neutrons with additional estimates made for those parents for whom a DS86 dose was not available. The cohort studied consisted of 31,150 live-born children, one or both of whose parents received >10 mSv, and two appropriate control groups totalling 41,066 children. Forty-nine malignant tumors were found in the control groups and 43 in the children of the exposed parents. A multiple linear regression analysis revealed no increase in malignancy in the children of the exposed parents. It was considered that 30 to 50 percent of the tumors of childhood were associated with an inherited genetic predisposition.

Recently it has been noted that the risk of leukemia in children whose fathers had been employed at the Sellafield nuclear plant in the United Kingdom was greater than in the control group (Gardner *et al.*, 1990). The highest relative risk, six- to eight-fold, was for children whose fathers were exposed to ≥100 mSv in the decade before the child's conception and four to five fold in those whose father's received ≥10 mSv during the six months before conception. The number of cases of childhood leukemia is small. The fathers of nine cases out of 46 cases of leukemia worked at the Sellafield plant. The relative risk of six (1.6 to 26.3) is based on four cases of leukemia in the children of the fathers who had received ≥100 mSv. The total number of controls for the leukemias in this case-control study was 488.

The results of this careful study must be confirmed or understood from a genetic and radiobiological basis before its importance and implications can be assessed fully. An examination of the genetic aspects concluded that, based on current knowledge, there is no theoretical basis to support the contention that the parental irradiation at the dose levels experienced at Sellafield could explain the incidence of leukemia reported (Abrahamson, 1990a).

Epidemiological studies of other exposed populations in the United Kingdom have failed to confirm an unequivocal association between paternal irradiation and the occurrence of childhood leukemia (McKinney *et al.*, 1991; Urquhart *et al.*, 1991). However, these studies did not have sufficient statistical power to detect an effect of comparable magnitude to that observed in the Sellafield study. The rates of mortality from childhood leukemia in United States counties containing nuclear installations have shown no excess (Jablon *et al.*, 1991), but studies specifically designed to examine the possibility of association between paternal irradiation and leukemia in their progeny would be required to eliminate any possibility.

9. Cancer and Alpha-Particle Emitting Radionuclides

9.1 Radon

The BEIR IV Report (NAS/NRC, 1988) derived lung-cancer-risk estimates for exposure to radon decay products based solely on epidemiological evidence. The committee used a descriptive analytical approach, obtained epidemiological data from four of the principal studies of radon-exposed miners (Ontario, Saskatchewan and Colorado Plateau uranium miners and Swedish metal miners), and developed modified multiplicative risk models for lung cancer based on analyses of these data. The database analyzed by the BEIR IV (NAS/NRC, 1988) contained a total of 360 lung-cancer deaths and 425,614 person years at risk. In these models, the excess relative risk varies with time since exposure, rather than remaining constant, and depends on age at time of exposure. Radon exposures more distant in time have a smaller impact on the age-specific excess relative risk than more recent exposures. In applying the model based on occupational data to the potential lung-cancer risk associated with indoor domestic exposure of the general population, the committee assumed that the epidemiological findings in the underground miners could be extended across the entire life span, that cigarette smoking and exposure to radon decay products interact multiplicatively, that exposure to radon progeny increases the risk of lung cancer in proportion to the sex-specific ambient cancer risk associated with other causes, and that a working level month (WLM) exposure[7]

---

[7]The traditional unit of exposure to radon progeny, the working level month (WLM), is used in this Report when referring to data from earlier publications rather than the SI unit of Jh m^{-3}. One WLM is equal to an exposure to one working level (WL) of short-lived radon progeny for a period of 170 hours, $i.e.$, one working month. One WL is equal to 1.3×10^5 MeV of alpha-particle energy emitted per liter of air from any combination of short-lived alpha-particle emitting progeny of ^{222}Rn, $i.e.$, primarily ^{218}Po and ^{214}Po. Therefore, the WLM is equal to the value of WL in the air being breathed times the number of hours of exposure divided by 170. An annual exposure of 1 WLM is approximately equal to a continuous annual exposure to a concentration of 4 pCi per liter of radon when the short-lived alpha-particle emitting radon progeny are in 50 percent equilibrium with the radon. In SI units 1 WLM is equal to 3.5×10^{-3} Jh m^{-3}.

yields a similar dose to the bronchial epithelium in both occupational and domestic environmental settings.

For lifetime exposure to 1 WLM y^{-1}, the lung-cancer risk is estimated to increase by a factor of 1.5 over the current rate for both males and females in the general population, assuming the current prevalence of cigarette smoking in the United States. Occupational exposure to 4 WLM y^{-1} from ages 20 to 40 y is projected to increase male lung-cancer deaths by a factor of 1.6 over the current rate in this age cohort in the general population. Much of the increased risk is in smokers for whom the risk of exposure to radon progeny is greater than in nonsmokers.

The model was used to project lifetime risks of lung cancer from lifetime exposure to radon progeny; an overall risk coefficient of 350 excess lung-cancer deaths per 10^6 person-WLM was obtained. A comparison of this lifetime lung-cancer-risk value with values derived in other studies (ICRP, 1987; NAS/NRC, 1980; NCRP, 1984; Puskin and Yang, 1988; UNSCEAR, 1977; 1988) is given in Table 9.1. These are summarized in the BEIR V Report (NAS/NRC, 1990) and as relative risks in a NCRP commentary (NCRP, 1991).

Direct comparisons of these risk estimates and the studies from which they are derived are not possible because of the differences in the models used, the populations assumed to be at risk (e.g., duration of exposure and smoking prevalence), differences in the assumed lung-cancer rates in the reference populations, and modeling of the smoking data and its interaction with alpha-particle radiation.

The UNSCEAR Report (UNSCEAR, 1988) recognized that the range of risk coefficients derived from various studies of uranium miners is broad, but in general compatible with the central value

TABLE 9.1—*Estimates of lifetime risk of lung cancer mortality due to lifetime exposure to radon progeny (adapted from Table 5-2, NAS/NRC, 1990).*

Source	Excess lifetime lung cancer mortality (deaths per 10^6 person-WLM)
UNSCEAR (1977)	200–450
UNSCEAR (1988)	150–450
BEIR III (NAS/NRC, 1980)	730
NCRP (1984)	130
ICRP (1987)	170–230[a]
	360[b]
Puskin and Yang (1988)	115–400
BEIR IV (NAS/NRC, 1988)	350

[a]Relative risk with ICRP (1987) reference population.
[b]Relative risk with 1980 the United States population used by BEIR IV (NAS/NRC, 1988).

of about ten excess lung cancers per 10^6 person-y WLM (additive risk model) or about a one percent increase in normal incidence of lung cancer per WLM (multiplicative risk model) (ICRP, 1987; UNSCEAR, 1988). When applied to the United States or Canadian male population, these risk coefficients suggest an average-lifetime risk of about three excess lung cancers per 10^4 person-WLM for uranium miners age 20 to 55 y at the time of exposure (Muller *et al.*, 1988). Recent data from those studies in which exposure data are reasonably reliable are comparable with the range of 1.5 to 4.5 excess lung cancers per 10^4 person-WLM for adult male miners.

The data in Table 9.1 of lifetime risk estimates of lung-cancer mortality suggests that the range of values is fairly broad; this is largely due to the difference in models used and the reference populations used to project lifetime risks. The analyses over the past decade indicate that the BEIR IV (NAS/NRC, 1988) estimate is near the middle of the range of risk estimates. It is about three times larger than the NCRP (1984) value, and about half of the estimate of the BEIR III (NAS/NRC, 1980). The latter two reports assumed an additive risk model; the BEIR III (NAS/NRC, 1980) based its projections on a model that is constant over time and on an increasing excess risk with age, while the NCRP (1984) projection is based on a risk model with diminishing excess risks with time after exposure. The BEIR IV (NAS/NRC, 1988) estimate is based on a time- and age-dependent modified multiplicative risk-projection model that takes into account the reduced risk at age 65 or greater and the small effectiveness of exposures occurring 15 y or more in the past, elements that were identified in the miner-cohort data. The U.S. Environmental Protection Agency (EPA) estimate is based on a constant multiplicative risk model and a United States reference population (Puskin and Yang, 1988). The ICRP (1987) estimate is based on a constant multiplicative risk-projection model and a European reference population; lifetable methods used by UNSCEAR (1988) are essentially identical to those used by ICRP (1987). When the BEIR IV (NAS/NRC, 1988) and ICRP values are calculated for the same 1980 United States reference population, the risk estimates are almost the same, *viz.*, 350 and 360 excess lung cancers per 10^6 person-WLM, respectively. The latest EPA estimate (Puskin and Yang, 1988) is based on an average of the BEIR IV (NAS/NRC, 1988) and ICRP (1987) values and the 1980 to 1984 United States reference populations. One reason the BEIR IV (NAS/NRC, 1988) risk-projection model was developed with more extensive modifications than the others, was that the original data from the epidemiological studies of the miners were made available to the committee for combined analyses, while other reports relied solely on published data that frequently lacked information on time- or age-dependent factors that modify risks (NAS/NRC, 1990). The UNSCEAR (1988) found no recent data or analyses that suggest any reason for a change in the previous lung-cancer-risk estimates of 1.5 to 4.5 fatal lung cancers per 10^4 person-WLM, *i.e.*, in accord with that estimated by BEIR IV (NAS/NRC, 1988).

The BEIR IV Committee (NAS/NRC, 1988) examined the comparative dosimetry of radon and radon progeny in mines and homes in relation to the dose of alpha-particle radiation to target cells in the respiratory epithelium. Differences in exposure-dose relationships in mines and in homes are expressed as a ratio, termed k in the BEIR IV Report (NAS/NRC, 1988); the ratio represents the quotient of the dose of alpha-particle energy delivered per unit exposure to the individual in the home to the dose per unit exposure to an individual miner. After examining a number of physical and biological parameters, k factors for bronchial dose were calculated for normal people, without respiratory illnesses, in the general environment. The committee's calculations indicate that the dose of alpha-particle energy per unit exposure delivered to the secretory and basal cells in the respiratory tract tends to be lower for the home environment; by about 30 percent for adults of both sexes and by 20 percent or less for infants and children. Thus, based on the BEIR IV Report (NAS/NRC, 1988), direct extrapolation of lung-cancer-risk estimates from mines to the home environment may overestimate the numbers of radon-caused lung-cancer cases by these percentages.

9.2 Radium

The main sources of information on the health effects of radium deposited in human tissues are the United States cases of occupational exposure (mainly radium dial painters and radium chemists) and medical exposure to ^{226}Ra or ^{228}Ra and the German experience with patients treated with ^{224}Ra for ankylosing spondylitis or tuberculosis. In the dose range in which bone tumors have occurred, the lifetime risk associated with ^{224}Ra was estimated in the BEIR IV Report (NAS/NRC, 1988) to be about 200 excess bone sarcomas per 10^4 person-Gy when a linear response is assumed. There was also an apparent increase in risk with dose protraction. There is considerable uncertainty about the risk using a life-table analysis. Mays and Spiess (1983) estimated the risk to juveniles to be 189 ± 32 per 10^4 person-Gy and to adults to be 133 ± 36 person-Gy. The analysis of

Chmelevsky *et al.* (1986) established that there was not a significant difference between the risks for juveniles and adults. Furthermore, they found that the dose-response relationship was not linear over the complete range of skeletal doses, but at very low doses the risk did vary linearly with dose (Figure 9.1). For ^{226}Ra and ^{228}Ra bone sarcoma induction in terms of estimated dose, the data are less clear, and risk estimates are uncertain.

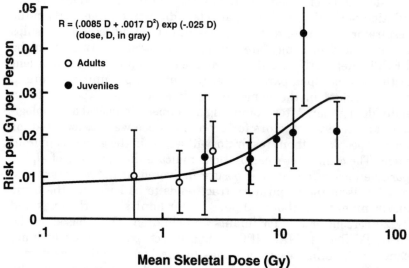

Fig. 9.1. The risk per person-Gy as a function of the mean skeletal dose estimated by the proportional hazard method of analysis. The risk ± standard errors are shown for adults: ○ and juveniles: ● (from Chmelevsky *et al.*, 1986).

Carcinomas in the paranasal sinuses and the cells in the mastoid air spaces are observed after exposure to ^{226}Ra or to ^{226}Ra combined with ^{228}Ra (see Table 9.2). The risk coefficient was estimated by BEIR IV (NAS/NRC, 1988) to be approximately 16 excess cancers per 10^6 person-y at risk per 3.7×10^4 Bq of intake. The UNSCEAR (1988) Report cited an absolute risk coefficient for bone cancer and alpha-particle emitting radionuclides of 200 per 10^4 person-Gy.

TABLE 9.2—*Carcinomas of the mastoid and paranasal sinuses in radium-dial painters exposed to ^{226,228}Ra (adapted from Table 4-4, NAS/NRC, 1988).*

Site	Number of cases
Mastoid	24
Ethmoid	2
Ethmoid/sphenoid	2
Sphenoid	6
Frontal	1
Total	35

9.3   Thorium

Risk estimates for ^{232}Th-induced liver cancer, bone cancer and leukemia have been calculated by BEIR IV (NAS/NRC, 1988) on the basis of the German thorotrast patient study (NAS/NRC, 1988; van Kaick *et al.*, 1989). For liver cancer, a lifetime risk is estimated to be about three excess cancers per 10^2 person-Gy, where the alpha-particle radiation dose is to the liver. UNSCEAR (1988) reviewed the thorotrast data and details from the German thorotrast series (van Kaick *et al.*, 1989). The lifetime risk estimate was three liver cancers per 10^2 person-Gy and the estimated risk rate coefficient was 0.13 excess liver cancers per 10^2 person-y Gy. For bone sarcomas, the lifetime risk is estimated by BEIR IV (NAS/NRC, 1988) to be about 0.6 to 1.2 excess cancers per 10^2 person-Gy, where the dose is to the skeleton only, without consideration of the dose to bone marrow. For leukemia, a lifetime risk was estimated to be about 50 to 60 per 10^4 person-Gy (NAS/NRC, 1988).

9.4   Transuranic Elements

In the absence of sufficient human epidemiological data, cancer-risk estimates for transuranic elements (*i.e.*, plutonium) are usually estimated on the basis of human studies of other alpha-particle emit-ting radionuclides (*e.g.*, lung cancer in underground miners exposed to radon progeny, bone cancer in radium-dial painters and patients exposed to radium) and of low-LET radiation exposures. The BEIR IV Report (NAS/NRC, 1988) estimated the lifetime risks of exposure to inhaled transuranic elements (plutonium) are about seven excess lung cancer deaths per 10^2 person-Gy, using a nominal value of 5 mGy per WLM of exposure. Based on a Bayesian analysis of animal and human data sets, the estimate of the risk of bone cancer following plutonium deposition in human bone is about three bone-cancer deaths per 10^2 person-Gy (NAS/NRC, 1988). For liver cancer, a risk estimate for plutonium is about 2.6 to 3.0 fatal liver cancers per 10^2 person-Gy. Caution was recommended in the application of these risk estimates to the transuranic elements. Their origins as well as the large uncertainties associated with their calculation should be taken into consideration. For example, the risk estimates depend on what latent period is assumed, the route of expression and the dose distribution.

9.5 Summary

The lifetime cancer-risk estimates following exposure to selected internally-deposited alpha-particle emitting radionuclides were summarized by BEIR IV (NAS/NRC, 1988) and are shown in Table 9.3.

TABLE 9.3—*Lifetime cancer-risk estimates from internally-deposited alpha-particle emitting radionuclides [data from BEIR IV (NAS/NRC, 1988)].*

Radionuclide	Tissue	Excess cancer deaths
^{222}Rn and radon progeny	Lung	350 per 10^6 person-WLM
^{224}Ra	Bone	200 per 10^4 person-Gy
^{226}Ra plus ^{228}Ra	Paranasal sinuses and mastoid air cells	16 per 10^6 person per y per 3.7×10^4 Bq
^{232}Th	Liver	300 per 10^4 person-Gy
	Bone	55–120 per 10^4 person-Gy
	Bone marrow	50–60 per 10^4 person-Gy
Plutonium	Lung	700 per 10^4 person-Gy
	Bone	300 per 10^4 person-Gy
	Liver	300 per 10^4 person-Gy

10. Iodine Radioisotopes and Thyroid Cancer

NCRP Report No. 80 (NCRP, 1985) provides estimates of risk of thyroid cancer and thyroid nodules following exposure to x rays. The data from the studies of children exposed to external x radiation can be used, with modification, in the calculation of the risk from the radioiodine internal emitters. The model used for calculation of the radiation risk for thyroid carcinogenesis from internally deposited radioiodine was:

$$\text{Risk} = R \times F \times S \times A \times Y \times L \qquad (10.1)$$

where: R is the absolute risk estimate (excess cases per 10^4 person-y Gy) for both sexes of children exposed to external x radiation and a minimum induction period of 5 y; F is the dose-effectiveness reduction factor (1.0 for external x radiation and ^{132}I, ^{133}I, ^{135}I and 0.33 for ^{131}I and ^{125}I); S is the sex factor (1.33 for females, 0.67 for males); A is the age factor (one for persons <18 y of age and 0.5 for persons >18 y); Y is the average number of years at risk; and L is the lethality factor (0.1 for maximum lifetime lethality).

This model was applied to the estimation of both the annual and lifetime risk of fatal thyroid cancer in the United States population. The estimates of the annual risk of fatal thyroid cancer in relation to sex and age at exposure are shown in Table 10.1. The projected lifetime risk of fatal thyroid cancer for the general population was estimated to be 7.5×10^{-4} Gy^{-1}.

TABLE 10.1—Annual risk of excess lethal cancers per 10^4 person-Gy of dose to the thyroid for doses from 0.06 to 15 $Gy^{a,b}$ (NCRP, 1985).

Source of Radiation	Age at exposure >18 y of age		Age at exposure ≤18 y of age	
	Female	Male	Female	Male
^{131}I, ^{125}I	0.028	0.056	0.056	0.112
External x or γ rays and ^{132}I, ^{133}I and ^{135}I	0.084	0.168	0.168	0.336

[a] United States population.
[b] Based on an absolute risk of 2.5 excess cases per 10^4 person-y Gy at risk in persons exposed to external radiation in childhood.

77

Following an extensive review of the human epidemiological and experimental laboratory data on the relative carcinogenicity of ^{131}I and external x radiation, NCRP (1985) concluded that ^{131}I was less carcinogenic in humans than radiation from external sources on the basis of estimated dose. Available human data on low-dose ^{131}I exposures suggest that the carcinogenic effect of ^{131}I in the human thyroid is low (Holm et al., 1988; NCRP, 1985; NRC, 1989; UNSCEAR, 1988). A comparison for children between the annual risk coefficient for ^{131}I of 0.6×10^{-4} person-y Gy and the risk of thyroid cancer of 2.4×10^{-4} person-y Gy following external x radiation suggests that ^{131}I is no more than about one-fourth as effective as external radiation (NCRP, 1985). For adults, a similar risk calculation of 0.7×10^{-4} person-y Gy for ^{131}I and 1.25×10^{-4} person-y Gy for external radiation, suggests that ^{131}I is no more than one-half as carcinogenic as external radiation. Based on human experience, the NCRP Report (NCRP, 1985) concluded the range of relative effectiveness of ^{131}I compared to external radiation is between 0.2 and 1. Combining the calculated effectiveness values in children (0.25) and adults (0.5) for application to the general population, an upper limit value of 0.33 is recommended for application to the general population, that is, the relative effectiveness of a given dose from ^{131}I compared to the same dose to the thyroid from external radiation is no more than one-third.

The U.S. Nuclear Regulatory Commission (NRC) Health Effects Models Report (NRC, 1989) applied linear risk coefficients for calculation of estimates of excess thyroid cancer and for benign thyroid nodules in the general population in the United States following exposure to the radioiodines. These are derived from several epidemiological studies on North American cohorts, and the coefficients were used to provide upper, central and lower estimates. The NRC report chose the NCRP (1985) risk estimate for determining lifetime risks. The pooled data of thyroid cancer following head and neck x-irradiation for benign disease in childhood in the North American cohorts comprised: 7,829 patients irradiated, 109 excess thyroid cancers, mean of 21.2 y at risk assuming a minimum induction period of 5 y, mean thyroid dose of 2.45 Gy and 429,149 person-y at risk. Since thyroid effects were observed mainly in populations who were very young at the time of exposure, and where follow-up periods included years when spontaneous risk was very low, the NRC (1989) committee concluded that it was difficult to estimate relative risk coefficients reliably. Furthermore, since spontaneous rates for thyroid cancer in the United States population show very little increase with age after about age 30 y, differences in lifetime risks based on relative and absolute risk-projection models are not appreciable. In general, risks of thyroid effects appear much smaller for those who are older at exposure, as there is very little evidence of radiation-induced thyroid effects for those exposed over 30 y of age. Risks are greater for females than for males, although in the general population consisting of equal numbers of males and females, the effects of sex-related differences on the total population risk would not be large. The committee used a lethality factor of 0.1, although it was recognized that the distribution over time is different for mortality and incidence, with deaths tending to occur later in life.

Based on epidemiological studies of thyroid carcinogenesis after therapeutic exposures to ^{131}I, the NRC (1989) committee concluded that there is no evidence of ^{131}I-induced thyroid carcinogenesis at high-dose levels (greater than 20 Gy) in adults. In studies of diagnostic exposures to ^{131}I, an excess risk was observed in ^{131}I exposed patients, but this was less than the risk following exposure to the same dose from external radiation. In a study in Sweden of about 35,000 patients administered ^{131}I for diagnostic purposes, the mean absorbed dose in the thyroid was estimated to be about 0.5 Gy. The patients have been followed for an average of 20 y and 50 cases of thyroid cancer were identified, compared to about 39 cases expected, based on the rate for the general population, the standardized incidence ratio was 1.27 (95 percent confidence interval 0.94 to 1.67). It was concluded that the risk of radiation-induced thyroid cancer might be about four times less from irradiation from the level of ^{131}I used in diagnosis (about 2 mBq) than from comparable levels of acute external x irradiation (Holm et al., 1988; 1989). Based on the human experience, the NRC committee (NRC, 1989) concluded ^{131}I was no more than one-third as carcinogenic to the thyroid as external x irradiation. Based on the animal data, the risk estimates for external radiation were multiplied by 0.1 (lower bound), 0.33 (central value) and 1.0 (upper bound) to give risk estimates for exposure to [131]I. For benign thyroid nodules, a lower bound is taken as 0.2 and an upper bound of 1.0 of those values for external radiation. Based on an absolute risk estimate of 9.3 cases of thyroid nodules per 10^4 person-y Gy in children exposed to external x irradiation in childhood, the risk estimates can be calculated (see Table 10.2).

TABLE 10.2—*Annual risk of excess benign thyroid nodules per 10^4 persons per Gy of thyroid dose from 0.06 to 15 Gy (United States population) (from Table A.7, NRC, 1989).*

Source of Irradiation	>18 y of age at exposure		<18 y of age at exposure	
	Male	Female	Male	Female
[131]I	0.6	1.2	1.2	2.5
External x or gamma rays	3.1	6.2	6.2	12.4

The BEIR V Report (NAS/NRC, 1990) analyzed the long-term follow-up of two study populations of children who were treated with x irradiation for benign conditions, *viz.*, the Israeli *tinea capitis* cohort (Ron and Modan, 1984) and the Rochester thymus cohort (Shore *et al.*, 1984). The committee applied an analysis of the data based on a dose-time-response modified linear model and derived risk-estimates based on a multiplicative-risk projection model. The relative carcinogenic potency of ^{131}I compared to x rays in producing thyroid cancer was estimated to be 0.66. The analysis of Laird (1987) provided the basis for the risk ratio for ^{131}I relative to x rays. The value of 0.66 (95 percent confidence limit, 0.14 to 3.15) had broad confidence limits, recognizing that the risks from radionuclides of iodine are not well understood. The committee's analyses of the data were carried out with the use of programs (AMFIT) originally developed for analyses of cancer mortality and incidence in the Japanese atomic-bomb LSS, with a minimum latent period of five years. The risk of thyroid-cancer under various modeling assumptions was calculated for males and females, ages 0 to 4 and 5 to 15 y at exposure and minimum latency intervals of 20 and 30 years. For the constant additive risk model, the range calculated was 2.3 to 25.5 thyroid-cancer cases per 10^4 person-y Gy at age 40 y; the constant relative risk model, for males and females, ages 5 to 15 y at exposure, yields a low estimate of 8.3 at 2 Gy at age 40 y, and for ages 0 to 4 y at exposure, a high of 23.6 cases per 10^4 person-y Gy at age 40 years. It was considered unlikely that risk-projections based on an excess-absolute-risk model could provide reliable indications of lifetime risk when applied to different populations. Accordingly, the committee chose to calculate projections of lifetime thyroid-cancer risks based on the preferred constant relative risk-projection model for Israeli-born children who were over 5 y at the time of irradiation. In this model, the relative risk at 1 Gy at age 40 y was 8.3 (95 percent confidence limit, 2 to 31) thyroid cancer cases. On the basis of unpublished data for atomic-bomb survivors, it was concluded that the risk estimate of radiation-induced thyroid cancer in adults is, at most, one-half that in children.

UNSCEAR (1988) reviewed NCRP (1985) and considered those results and analyses to provide the best estimates of thyroid-cancer risks. The specific risk estimate model was adopted so that the risk could be calculated for incidence and mortality. The committee noted that NCRP (1985) compared absolute and relative risk models and found little difference in lifetime estimates. In view of the concordance of views, the ICRP (1991) adopted a lifetime-risk estimate for fatal thyroid cancers of 7.5×10^{-4} Gy^{-1}. This value was estimated for the high-dose range, and it is not modified downwards because of the presumed linear nature of the thyroid-dose response. ICRP (1991) proposed a thyroid cancer lethality fraction of 0.10 in its estimation of thyroid-cancer risk, and a thyroid tissue weighting factor (w_T) of 0.05 for both sexes and a wide range of ages.

The estimates of the lifetime risk of fatal thyroid cancer by the various committees since 1977 are shown in Table 10.3.

In this Report, the NCRP has again reviewed the risk estimates for thyroid cancer and benign nodules and finds the most reliable values to be those reported in NCRP Report No. 80 (NCRP, 1985) and the NRC Health Effects Models Committee (NRC, 1989). Based on an absolute risk estimate of 2.5 cases per 10^4 person-y Gy in children exposed to external x irradiation in childhood, the risk coefficients that are considered applicable to the general population of the United States for mean thyroid doses ranging from 0.06 to 15 Gy, depend strongly on age and sex. Exposure before the age of 5 y may result in about a three-fold greater increase in thyroid cancer than exposures at older ages. Females are two to three times more susceptible to both naturally occurring and radiation-induced thyroid cancers. If these risk coefficients are applied to the general population, then the lifetime incidence of fatal thyroid cancer would be seven to eight cases per 10^4 person Gy following exposure to external x irradiation for a population comprised of equal proportions of males and females and of adults and children. Although the data were derived from persons exposed to high-dose-rate irradiation, no DREF is applied. Approximately ten percent of the thyroid cancers will be lethal. Until further data become available, ^{131}I is considered to be one-third as effective as external x radiation in the induction of thyroid cancer in the general population. The absolute risk of benign thyroid nodules following external radiation therapy in childhood is estimated to be 10.3 cases per 10^4 person-y Gy.

TABLE 10.3—*Lifetime fatal thyroid-cancer-risk estimates.*

Report	Risk estimate (cases per 10^4 person-y Gy)
UNSCEAR (1977)	5–15
ICRP (1977)	5
BEIR III (NAS/NRC, 1980)	6–18
NCRP (1985)	7.5
UNSCEAR (1988)	7.5[a]
NRC (1989)	7.5[a]
BEIR V (NAS/NRC, 1990)	8.3–23.6 relative risk at 1 Gy[b]
ICRP (1991)	8.0

[a]Based on the NCRP (1985) risks estimates.

[b]Absolute risk of 2.3 to 25.5 excess cases per 10^4 person-y Gy at age 40 years.

11. Hereditary Effects

11.1 Introduction

The past 35 y of genetics research has led to a major and continuing revolution in our knowledge of the structure of the genetic material, the organization of genes of higher organisms and very precise knowledge of the structure of specific human genes and their functioning. As yet, the new knowledge has not altered the approach to the estimation of the risks of genetic effects induced by ionizing radiation (Neel *et al.*, 1990). This is not to say that there have been no molecular inroads into studies of genetic effects of ionizing irradiation, but the impact to date has been small when viewed from the perspective of induced genetic disease.

Unlike somatic health effects induced by radiation, for which extensive epidemiological evidence exists demonstrating a dose-response relationship for leukemia and solid cancers, no similar body of positive dose-response data exists for purposes of human genetic risk estimates. The large scale human genetic studies carried out to-date have shown no significant increase with increasing dose in the measured genetic end points. These studies will be discussed in greater detail subsequently. But, these human studies do not stand alone, there have been nearly seven decades of radiation genetics research studies on animals and plants, demonstrating that radiation induces gene mutations, chromosome rearrangements and losses or gains of whole chromosomes or large segments within them, and in a dose-related manner. Therefore, there can be little doubt that humans respond in a somewhat similar fashion.

The issue then becomes how best to prudently estimate the genetic risks for humans. This issue entails the following elements:

1. The genetic bases for the variety of diseases affecting humans, their incidence and or spontaneous mutation rates
2. The germ cell stages of greatest importance for the induction of mutations and use of an appropriate animal surrogate when necessary
3. The radiosensitivity of measurable and relevant genetic end points in a surrogate system
4. Extrapolation procedures for interspecies analysis

82

11.2 Estimates of Genetic Risk

The doubling dose (DD) is that dose of radiation (or any mutagen) which will double the spontaneous mutation rate in a biological system. It follows that the larger the DD the less radiosensitive the system. The DD can be very accurately estimated when very specific mutational end points are examined, such as recessive visible or dominant gene mutations at specific loci, or chromosome translocation rates. Once the induced mutation rate is determined for a given set of doses which demonstrate a linear-dose-response relationship, the DD is computed by dividing the spontaneous mutation rate by the induced rate for the same end points.

In applying the relative mutation risks (1/DD) to the natural incidence (per million live born) of a class of genetic diseases, such as dominant disorders, the value initially derived is the total number of cases to be expected over the equilibrium time, in n generations, through which the mutant genes will persist. The natural incidence is composed of mutants arising in some number of previous generations with an average persistence time until elimination. For most dominant disorders that time is estimated to be about five to six generations and the estimate of first-generation effects is one-fifth to one-sixth the equilibrium estimate. Therefore, unlike induced somatic effects which only affect the exposed population, genetic risk estimates attempt to deal with the total number of disorders that will be introduced, not only to the next generation of offspring, but to successive generations as well. It has been the practice of the genetics committees to present their calculations in terms of first-generation and equilibrium-generation estimates, *i.e.*, the total number of cases expected to occur per million live born. Societal judgments will usually be based on the estimates of risks for the first generation only.

Risk estimates for the spermatogonial stage in males and for the immature oocytes of females are considered most important for the estimation of the DD for humans, because the cells in these stages accumulate mutational damage throughout the reproductive lifetime of the individual.

Estimating DD for humans is more difficult than for test organisms for many reasons which will be considered later. Until recently, estimates of the human DD were derived from studies on mouse germinal gene mutation rates with the assumption that they are representative for the human.

The approaches of both the UNSCEAR (1988) and BEIR V (NAS/ NRC, 1990) to the estimation of radiation-induced genetic risks and the components of that risk are considered together and are compared to other estimates. The BEIR V (NAS/NRC, 1990) employed a DD of 1 Sv to derive values for first-generation dominant disorders and congenital effects as well as the equilibrium values for these end points. The derivations of the DD or its inverse, the relative mutation risk per unit dose, are described in detail in the BEIR V Report (NAS/NRC, 1990) and permit their calculation given the necessary assumptions. The fact that a DD of 1 Sv is used by the BEIR V (NAS/NRC, 1990) is, in itself, not surprising; this agrees with the mouse-derived DD employed by the UNSCEAR (1988) as well as past reviews. The BEIR I (NAS/NRC, 1972) and BEIR III (NAS/NRC, 1980) employed DD values of 0.2 to 2.0 Sv and 0.5 to 2.5 Sv, respectively, based on induced rates in mice, and human or mouse spontaneous rates. The significant aspect of the DD in the BEIR III Report (NAS/NRC, 1980) is that it derived estimates from the lower 95 percent confidence limit for malformations, stillbirth and neonatal deaths, end points observed in the studies at the RERF of the offspring of parents who survived the Hiroshima and Nagasaki atomic bombs. It should be noted, as mentioned earlier, that studies of the Japanese have not shown a significant dose-related increase in any of these untoward pregnancy outcomes. The BEIR V (NAS/NRC, 1990) literally rounded off the values presented by Schull *et al.* (1981) to the 1 Sv value. It should also be noted that the earlier estimates were based on the T65DR which has now been superseded by the DS86 which estimates lower total doses and, specifically, considerably lower contributions of neutrons to the total dose. How this would have affected the lower 95 percent estimates is unclear, but if not dramatically, the estimate of 1 Sv would have been kept, if only to make the calculations easier and to demonstrate that no great precision was intended. The recently developed central estimates of the DD based on the data from the study of the progeny of the atomic-bomb survivors (Neel *et al.*, 1990) are, however, considerably higher, a factor of three or more than that derived from the seven specific locus studies in the mouse (NAS/NRC, 1972; 1980). There are, however, major differences in the end points on which the human estimates are based when compared to the specific locus studies in the mouse.

11.3 Doubling Dose and Dominant Diseases

BEIR V (NAS/NRC, 1990) has introduced new approaches to the estimation of dominantly inherited genetic diseases resulting from radiation. The current incidence, 10,000 cases per million live born, is the same as used by the BEIR III (NAS/NRC, 1980) and UNSCEAR (1988). However, this was further subdivided into those which would be recognized as clinically severe disorders, and those described as clinically mild. The incidence of the severe disorders is 2,500 cases per 10^6 live born and with a selective disadvantage of 20 to 80 percent that of normal. The greater the selective disadvantage the lower the survival and the fewer number of generations the disease will persist in the population. The incidence of the "clinically mild" diseases is 7,500 cases per 10^6 live born. These disease cases have a 1 to 20 percent selective disadvantage compared to normals and thus would have a longer persistence time to equilibrium. Therefore, the BEIR V Report (NAS/NRC, 1990) estimated that a total of between 6 to 35 cases per million would be expected in the first generation following an additional 10 mSv parental exposure. The estimates of BEIR III (NAS/NRC, 1980) range from 5 to 65 cases, based on the dominant mouse skeletal mutation rates, the so-called direct estimate. The BEIR III (NAS/NRC, 1980) values were also based on a different mutation contribution. The contribution of females was taken to be 44 percent of that from males, based on other mouse experiments. The BEIR V Report (NAS/NRC, 1990) circumvents this issue with the DD method it employs and implicitly assumes an equal contribution of mutation in the male and female gametes for this end point as well as all other end points.

The UNSCEAR (1988) and previous reports have derived first-generation effects using separate assumptions for the male and female germ cell contributions, again assuming lower contribution from females for both dominant diseases and diseases resulting from structural chromosome rearrangements. The UNSCEAR (1988) estimated a range of 10 to 29 cases of dominant diseases in the first generation with a single point estimate of 15 cases, the latter value based on a DD of 1 Sv. The 10 to 29 cases are estimated differently using the dominant mutation rates of mouse cataracts and skeletal defects to compute "directly" the first-generation total dominant disorder frequency. Entailed in these calculations are a large number of assumptions and extrapolations which have been subject to some controversy. The BEIR V Report (NAS/NRC, 1990) stated that it "had little confidence in the reliability of the individual assumptions required by the direct method let alone the product of a long chain of uncertain estimates that follow from these assumptions."

X-linked recessive deleterious mutations result in diseases that affect males more frequently than females because males have only one X chromosome, which allows the immediate expression of a recessive gene, whereas females have two X chromosomes and mutant expression requires homozygosity of the recessive genes. The

BEIR V (NAS/NRC, 1990) applied DD methodology to derive its estimate of less than one induced case per 10^6 per 10 mSv parental exposure. A different approach was employed by NRC (1989), which assumed there could be as many as 250 genes on the X chromosome that could result in serious disease consequence. This number was multiplied by the specific locus rate, *i.e.*, 7.2×10^{-8} mutations per locus per 10 mSv, derived from the mouse studies and corrected for those mutations that would have been maternally derived. This produced a value of about eight first-generation cases. There would have to be less than 30 genes on the X chromosome at the given induced mutation rate, to produce the BEIR V (NAS/NRC, 1990) estimate. Thus two values (<1 to 8) probably provide a reasonable lower- and upper-bound estimate of induced X-linked mutational disease for the stated conditions.

In summary, for human diseases of a dominantly inherited nature, the BEIR V (NAS/NRC, 1990) estimates a range from 6 to 35 first-generation cases per 10^6 live born per 10 mSv while UNSCEAR's (1988) estimate of 15 cases is the geometric mean of the BEIR V (NAS/NRC, 1990) estimates. Both sets of estimates were derived by employing the same DD estimates. First-generation X-linked diseases were estimated at less than one case and eight cases in the BEIR V Report (NAS/NRC, 1990) and the NUREG (NRC, 1989) analysis, respectively, while UNSCEAR's (1988) estimate is included within the dominant cases.

11.4 Aneuploidy Induction

There are presently estimated to be about 3,800 aneuploid births per 10^6 live born. The vast majority of these chromosome abnormality cases do not reproduce, so that the equilibrium time is effectively one generation.

The issue of whether irradiation will increase the rate of nondisjunction, the improper segregation of chromosome pairs during meiosis, leading to an increase in aneuploid births (individuals with abnormal chromosome numbers, 45 or 47, instead of the normal diploid 46 chromosome number) has received considerable attention. Particularly since cytogenetic studies demonstrated the aneuploid nature of Down, Klinefelter's and Turner's syndromes and other more lethal types of chromosome abnormalities in live born and abortuses. The BEIR V Report (NAS/NRC, 1990) reviewed past studies in this area and concluded that "the induction of nondisjunction by low-level irradiation of immature oocytes may not present a serious concern." The report presents a value of <1 additional case per 10^6 live-born offspring per 10 mSv per generation. This is evidently not a calculated number derived from experimental studies, but is apparently the committee's upper estimate of a nonsignificant effect. In effect, the DD estimate is 4.0 Sv at a minimum. On the other hand, had the committee used the lower 95 percent confidence limits of the 1981 atomic-bomb aneuploidy data (as was done for dominant disease estimates), they would have derived a larger risk value, namely 12 cases per 10^6 live born per 10 mSv and a DD of about 3 Sv. This high DD value suggests that none of the earlier animal experiments attempting to detect induced aneuploidy were likely to provide positive results given the size and conditions of the experiments. Moreover, the most recent estimates (Neel et al., 1990) would suggest a DD in the range of 6 to 7 Sv with a lower 95 percent confidence limit (adjusted for dose rate) of about 4 Sv.

The UNSCEAR Report (UNSCEAR, 1988) also provides no recent estimates on aneuploidy induction. An important paper by Griffin and Tease (1988) on the induction of anomalies in mouse immature oocytes by low-dose-rate gamma rays was not considered in either the BEIR V (NAS/NRC, 1990) or UNSCEAR (1988) Reports. Also, Martin et al. (1986) have reported on induced nondisjunction in human spermatogonial cells and demonstrated that among the sperm derived from these irradiated stem cells, a substantial proportion were aneuploid; moreover, the response was dose dependent. Using the data from the Griffin and Tease (1988) and Martin et al. (1989), a very conservative set of assumptions (Abrahamson et al., 1990b; 1990c), leads to an estimate of nine cases per 10^6 live born per 10 mSv, which suggest a DD in the range of 4 Sv.

Given the newer cytogenetic techniques, such as "painted chromosomes," fluorescence in situ hybridization of specific chromosomes and chromosome centromeres, in combination with electronic scanning equipment, one would expect that critical experimental animal studies could provide reasonable answers.

11.5 Translocations

Translocations represent the second class of microscopically observable chromosome abnormalities, which result from breakage and structured reorganization of parts of different chromosomes. They are called balanced or reciprocal translocations when the reorganization, usually between two nonidentical chromosomes, does not lead to an unbalance of genetic material or to acentric or dicentric chromosomes. An individual inheriting a reciprocal translocation may or may not suffer from a genetic disorder depending on whether there has been a disruption in genic activity at the sites of breakage. More important, however, in terms of genetic disorders is the probability that the meiotic products of a translocation-bearing germ cell will produce mature reproductive cells with unbalanced translocations, that is where only one-half of the reciprocal translocation is now present. In most cases, this genetic imbalance will result in zygotic inviability, because too great an amount of genetic information is lost. But in a small percentage of cases, grossly abnormal children are born but rarely survive infancy. Those individuals receiving a reciprocal translocation do reproduce and their offspring may inherit the unbalanced translocation. On average, this type of chromosome abnormality has about a three-generation equilibrium time.

The BEIR V Report (NAS/NRC, 1990) concluded for translocations that its "review of the relevant data suggests that a value of 2×10^{-4}/rem would be more appropriate" and from this derived an estimate of less than five cases per 10^6 live born per 10 mSv of unbalanced first-generation diseases. This value is somewhat lower than that developed by the BEIR III (NAS/NRC, 1980) (<10 cases per 10^6 per 10 mSv). The computations by which the BEIR V (NAS/NRC, 1990) developed its estimate were not presented in the report; whereas reasonably detailed explanations were given for the derivations in the BEIR I (NAS/NRC, 1972), BEIR III (NAS/NRC, 1980), NRC (1989) and the UNSCEAR (1988) Reports and intercomparisons of the different assumptions by these other committees were also presented. The BEIR V Report (NAS/NRC, 1990), however, discusses the translocation induction rates in mice, and several species of primates including humans in a special review section on the mouse and other laboratory animals. Presumably, they developed their conclusion by using an induction rate, lower by a factor of three and a greater dose-rate reduction factor. But one cannot determine what intermediate assumptions they chose. These assumptions should include segregation factors for each sex and a viability estimate of the unbalanced products. The BEIR III Report (NAS/NRC, 1980) used a five percent viability factor. UNSCEAR (1986) employed a nine percent viability factor rather than the six percent value in its 1982 report. The UNSCEAR (1988) also derived different estimates for paternal and maternal exposures. The BEIR V Report (NAS/NRC 1990) provides no guidance on this issue. In conclusion, the BEIR V (NAS/NRC, 1990) unbalanced translocation estimate is less than five cases per 10^6 per 10 mSv. The UNSCEAR (1988) estimate is 1 to 15 cases from paternal exposure and zero to five from maternal exposure (using a mouse sensitivity model). The NRC report (NRC, 1989) estimated about 13 cases assuming equal sensitivity for both sexes but different segregation factors during meiosis.

11.6 Irregularly Inherited Diseases

Irregularly inherited disorder or traits also known as multifactorial disorders and complexly inherited diseases are those for which a genetic component is considered likely or has been established, but does not follow simple Mendelian inheritance. A significant fraction of the total genetic burden in humans consist of these traits, but because of the uncertainty of how they are inherited, there has been a reluctance to estimate an increase that might result from irradiation. In the 1977 report, UNSCEAR estimated the risk from multifactorial diseases based on the assumption of a prevalence of nine percent and a mutation component of five percent. In the 1988 report, an estimate was not made because of the uncertainties. The BEIR III Report (NAS/NRC, 1980) accepted the estimate of nine percent of genetic disorders of complex etiology and combined it with their DD exposure range of 1.29×10^{-2} C kg^{-1} to 6.45×10^{-2} C kg^{-1}, and a mutation component of 5 to 50 percent, to derive an equilibrium excess of 20 to 900 radiation-induced cases of irregularly inherited disease per 2.58×10^{-4} C kg^{-1} per 10^{6} live born.

It is important to examine the evidence for including risk coefficients for multifactorial disorders in the total risk estimates for genetic effects and to estimate their possible contribution. This is particularly the case since the UNSCEAR (1977) and BEIR III (NAS/NRC, 1980) Reports included this incompletely understood category of genetic effects whereas neither UNSCEAR (1988) nor BEIR V (NAS/NRC, 1990) did so. The BEIR V Report (NAS/NRC, 1990) states with regard to irregularly inherited diseases: "the committee has not made quantitative risk estimates. While, for this category of traits, the risks could be negligible, they could also be as large or larger than all the other traits combined" (see Table 2-1 of NAS/NRC, 1990).

The most important change in the estimation of the current incidence of genetic disease involves the complexly inherited diseases, excluding congenital anomalies. The change results from the introduction of three classes of diseases for which the BEIR V Report (NAS/NRC, 1990) assumes there is a genetic component, albeit not one for which a simple increase in mutation rate is likely to alter the incidence in a recognizable manner. The incidence of the three classes in the United States population are: heart disease, 600,000 cases per million; cancer, 300,000 cases per million; and other selected diseases, 300,000 cases per million; the latter value derived from recent studies on the prevalence of diseases in the population of Hungary and reported by UNSCEAR (1988) from the data of Czeizel et al. (1988). Thus, a total of 1.2×10^6 cases of diseases of these types will affect a population of 1×10^6 at some time during lifetime, most often after maturity. The value assumed in the BEIR III Report (NAS/NRC, 1980), which included congenital abnormalities as well, was 9×10^4 cases per 10^6 individuals. The discussion in BEIR V (NAS/NRC, 1990) deals with the lack of understanding of what the mutational component may be for either individual diseases or in total. The mutational component is that proportion of the disease incidence which will respond to a change in mutation frequency. When a mutational component is one (as in the case of dominants) then every new mutation will be expressed. For the complexly inherited diseases, the mutation component is most often not known and previous committees [BEIR I (NAS/NRC, 1972) and BEIR III (NAS/NRC, 1980)] assumed it would range between 5 and 50 percent. The BEIR V Report (NAS/NRC, 1990) provided discussion of the genetic conditions which determine a high, low or indeterminate mutational component and concluded that the mutational component is not known "even to its order of magnitude." Thus, this great uncertainty in the mutational component, as well as the equilibrium time for such diseases, lead to the committee's unwillingness to derive a risk estimate and to their conclusion as cited at the beginning of this Section. The UNSCEAR (1988) also did not develop a risk estimate for this end point.

A worst case analysis, using a mutational component of 0.5 and an equilibrium time of ten generations and a DD of 1 Sv for these diseases would lead to an estimate of 600 additional first-generation cases per 10^6 live born per 10 mSv of parental exposure. Assuming, at the other extreme, that the mutational component was 0.05 and an equilibrium time of 100 generations, the expected number would be six additional first-generation cases. The geometric mean of these two estimates is 60 cases of disease in the first generation per 10^6 live born per 10 mSv parental exposure.

For the component of genetic disease which results in congenital birth abnormalities (anencephaly, cleft palate, structural cardiac defects, etc.) about 20,000 to 30,000 cases per 10^6 births constitute the current incidence. For this group, the BEIR V Report (NAS/NRC, 1990) derives mutational components ranging from 0.05 to 0.35 percent, employing information such as trait heritability and monozygotic twin concordance to derive these values. Thus, incidence times the reciprocal of DD times the mutational component gives 10 to 100 cases per 10 mSv at equilibrium. A "worst case" assumption suggests about ten cases in the first generation if they were to result from dominant mutations.

While it is very unlikely that all these diseases have a dominant basis, for those that do, a mutational component of one is more appropriate, which could then raise their first-generation estimate by as much as a factor of three, *i.e.*, to perhaps as many as 30 first-generation cases.

The atomic-bomb genetics studies indicate a small but not significant increase in congenital anomalies and untoward pregnancy outcomes. Given that the mean parental dose was 0.18 Sv and some 12,400 children born to exposed parents were examined for congenital defects, the estimates shown above predict that between two and six additional cases might have been produced.

11.7   Risk Estimates: Atomic-Bomb Studies

Neel *et al.* (1990) have summarized their major studies on the genetic effects induced by the atomic bombs in the Hiroshima and Nagasaki survivors. This study has been continuing since 1948.[8] A fixed cohort of children born to either exposed or unexposed parents have been examined and followed for a variety of genetic end points. Those 76,617 children born through 1954 represent the subgroup for the indicator, untoward pregnancy outcomes. Approximately 13,000 children were born to exposed parents. Five sets of data were collected for analysis:

1. Untoward pregnancy outcomes, which include all stillborn, deaths within first two weeks and major congenital malformation
2. F_1 mortality, the survival of all children born between 1946 and 1985 was followed
3. Mutations affecting protein change, representing studies on 30 serum and erythrocyte proteins in over 23,000 children, 13,052 born to exposed parents
4. Over 16,000 children were examined cytologically for the presence of sex-chromosome aneuploidy, approximately half of the group were born to exposed parents
5. Data for malignancies in 31,150 children of exposed parents and 41,066 children of unexposed parents up to the age of 20 were analyzed

[8]The study included children born after May 1946.

While other end points were also analyzed, these five sets of data were used to compute dose-response relationships and regression coefficients. As stated earlier, none of these end points was significantly increased in the exposed group, and the regression coefficients were positive for three end points and negative for two others. For each end point, the authors estimate the contribution of spontaneous mutation, adjusting the background frequency by their best estimates of the respective mutational component. The sum of the spontaneous mutation rates for all end points, 0.0063 to 0.0084, was divided by the sum of the regression coefficients Sv^{-1}, 0.00375, to provide their best estimate of the DD of 1.7 to 2.2 Sv. Since the radiation was delivered instantaneously, these estimates were multiplied by a factor of two to adjust for a dose-rate reduction to chronic exposure. Their conclusion then is that the DD for those genetic disorders that lead to human morbidity and mortality is about 3.4 to 4.4 Sv for low-LET, low-dose-rate irradiation. This value is about three to four times higher than the DD value extrapolated from the mouse specific locus data under chronic exposure conditions.

Does this mean that the human species is indeed considerably less sensitive to irradiation than the mouse? There are considerable uncertainties in both estimates. In the estimates derived from the mouse, the extrapolation from one species to another based on a limited sample of genes, and in the estimates based on humans, there are reasonably large dose uncertainties that underpin the genetic assumptions. It is also necessary to recognize that the human DD estimates attempt to establish the total level of morbidity and mortality. Until further resolution of these uncertainties is reached, a reasonable course of action is to err on the conservative side and apply the higher risk estimates for population protection.

11.8 Summary of Risk Estimates

In Table 11.1, a composite of risk estimates for the end points discussed previously is provided. We use the BEIR V (NAS/NRC, 1990) estimate of the current incidence of diseases with a genetic component and provide a lower and upper estimate for all first-generation additional cases per 10^6 live born. For example, the additional cases per 10^6 live born is the incidence $\times$ 1/DD $\times$ selective disadvantage. In the case of the clinically severe autosomal dominant effects, the calculation is $2,500 \times 100^{-1} \times 0.2$ to $0.8 = 5$ to 20 and in the case of clinically mild effects, $7,500 \times 100^{-1} \times 0.01$ to $0.2 = 1$ to 15. The value presented for that estimate is the geometric

TABLE 11.1—*Estimates of genetic effects resulting from parental exposure of 10 mSv.*

Type of disorder	Current Incidence per 10^6 live born[a]	Additional cases per 10^6 live born	
		First Generation	Equilibrium
Autosomal dominant			
Clinically severe	2,500	5–20[a]	25
Clinically mild	7,500	1–15[a]	75
X linked	400	<1–8[b]	40
Chromosomal			
Unbalanced translocations	600	5[a]–20[b]	Little increase
Trisomies	3,800	<1–9[c]	Little increase
Irregularly inherited	1,200,000	60[d]	1,900[d]
Congenital abnormalities	20,000–30,000	10[a]–30[b]	10–100
Total	1,245,000	85–160[e]	2,050–2,140[e]

[a]From BEIR V (NAS/NRC, 1990).

[b]These values incorporate the range of estimates derived by BEIR V (NAS/NRC, 1990), UNSCEAR (1988) and NRC (1989; 1991).

[c]These include values derived in this Report.

[d]It should be noted that neither BEIR V (NAS/NRC, 1990) nor UNSCEAR (1988) attempted risk estimates for this end point because of the great uncertainty (see discussion).

[e]The totals are rounded-off values to indicate lack of precision.

mean of the wide range that can be predicted. Thus, the total first-generation effects are larger than those presented by either BEIR V (NAS/NRC, 1990) or UNSCEAR (1988). Table 11.2 provides the evolving estimates of those committees over the last two decades of effort. It should be noted that ICRP (1991) estimated the risk for serious hereditary effects for the total population to be $1.0 \times 10^{-2} \text{ Sv}^{-1}$ which included an estimate of the multifactorial disease component.

TABLE 11.2—*UNSCEAR and BEIR estimates of first-generation effects per 10^6 live born, per Sv of parental exposure.*

Type of Disorder	UNSCEAR committees			BEIR committees		
	1977	1982	1986–88	1972	1980	1990
Dominant + X linked	20	15	15	10–100	5–65	6–35
Chromosomal	38	2–20	2–20	10	<10	<6
Irregularly inherited	5	5	—[a]	1–100	—[b]	10
Total	65	20–40	20–35	20–200	15–70	20–50
Committee's natural incidence values	10.5%	10.5%	68%	6%	10.7%	125%

[a]No estimate produced because of great uncertainty; BEIR V (NAS/NRC, 1990) value of ten represents congenital abnormality estimate only.

[b]Cases expected were part of dominant estimate.

12. Radiation Effects on Brain

12.1 Radiation Effects on the Brain of the Embryo/Fetus

Data that are now available on the developmental anatomy of the mammalian brain have aided in the interpretation of radiation effects observed among the Japanese exposed *in utero* during the atomic bombings. Analyses of the atomic-bomb survivor data on those exposed *in utero*, together with the DS86, have permitted examination of the time-specific susceptibility to radiation-induced severe mental retardation, the most important developmental abnormality to appear in humans exposed prenatally, and has allowed the risk of these effects to be estimated. These findings have been reviewed in UNSCEAR (1988) and BEIR V (NAS/NRC, 1990). Both committees have used the same data bases, *i.e.* the studies of the children exposed *in utero* in Hiroshima and Nagasaki.

The studies of the atomic-bomb survivors indicate that there is an increase in the prevalence of mental retardation and small head size with increasing exposure (Blot and Miller, 1973; Miller and Blot, 1972; Miller and Mulvihill, 1976). In recent studies (Kriegel *et al.*, 1986; Otake *et al.*, 1988a; Schull and Otake, 1987), based on a cohort of 1,598 individuals, there were 30 children with severe mental retardation that were diagnosed before 17 years of age. In five of the subjects, there was a clinical history that presumably was nonradiation related that could account for their severe mental retardation; namely, three cases of Down syndrome, one with Japanese encephalitis in infancy, while another had a retarded sibling. Data analyses were made with and without these five cases. Estimates of fetal doses are not yet available from DS86 (Roesch, 1987), but DS86 organ dose estimates have been computed for most of the exposed mothers, and the uterine doses may be used. Four categories of gestational age were measured from the time of fertilization; these were 0 to 7, 8 to 15, 16 to 25 and ≥26 weeks. During the second period (8 to 15 weeks), there is a rapid increase in the number of neurons; they migrate to their developmental sites in the brain and lose their capacity to divide. A dose-dependent increased risk of severe mental retardation was observed in the gestational age group 8 to 15 weeks after fertilization (Figure 12.1) and, to a lesser extent, in the gestational age group 16 to 25 weeks after fertilization. In the age groups less than 8 weeks or greater than 26 weeks, no exposed subjects were mentally retarded. The relative risk for exposures during the 8 to 15 week period is four or more times greater than that for exposure during the 16 to 25 week period.

The data do not allow the precise definition of the dose response for the gestation periods 8 to 15 weeks and 16 to 25 weeks where there is an increase in severe mental retardation. Within these periods, a linear-dose response yields probabilities of induction per unit absorbed dose of 0.4 per Gy in the 8 to 15 week group and 0.1 per Gy in the 16 to 25 week group. Simple quadratic and linear-quadratic functions also provide acceptable fits to the data in both groups, particularly with DS86. There is also support for a dose threshold in the range of 0.1 to 0.2 Gy in the 8 to 15 week group and at about 0.6 to 0.7 Gy in the 16 to 25 week group. Exclusion of cases with possible nonradiation etiology exposed during the 8 to 15 week period (two cases with Down syndrome) increases the probability of a threshold. The threshold is 0.39 Gy for ungrouped data and 0.46 Gy when the data are stratified by dose interval. Both of these thresholds are significantly different from zero (lower 95 percent confidence limits, 0.12 Gy and 0.23 Gy, respectively).

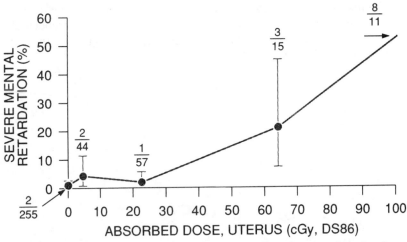

Fig. 12.1. Frequency of severe mental retardation, as a function of uterine dose, in children irradiated at gestational age between 8 to 15 weeks. Data from Hiroshima and Nagasaki are pooled, but the cases of Down syndrome are excluded (from Otake *et al.*, 1988a).

12.2 Radiation Effects on Intelligence

Intelligence-test (*Koga*) scores of Japanese atomic-bomb survivors of 10 to 11 y of age, who were exposed *in utero*, have been analyzed using estimates of the uterine absorbed dose based on the DS86 (Schull *et al.*, 1988). Children exposed at 8 to 15 weeks after fertilization, and to a lesser extent, those exposed 16 to 25 weeks after fertilization, show a progressive shift downward in individual scores with increasing dose. Based on a linear-dose-effect relationship, the diminution in intelligence score is 21 to 29 points at 1 Gy. Similarly, potential damage to the fetal brain in the 8 to 15 week group appears linearly related to fetal absorbed dose, based on simple regression of school-performance score on dose (Otake *et al.*, 1988b; Schull *et al.*, 1988).

12.3 Studies of Radiation-Therapy Patients

In the New York *tinea capitis* survey (Omran *et al.*, 1978; Shore *et al.*, 1976), there is a higher incidence of behavioral disorders and mental disease in the x-ray-exposed cohort of children treated for ringworm of the scalp. Brain doses were of the order of 1 Gy. In the Israel *tinea capitis* study (Ron *et al.*, 1982), the x-irradiated children have lower examination scores on scholastic aptitude, IQ and psychological tests, complete fewer school grade levels, and have a higher incidence of neuropsychiatric disorders. Children with acute lymphatic leukemia treated with irradiation to the brain are reported to have lower IQ scores and display disturbances in cognitive functions (Meadows *et al.*, 1981).

13. The Relative Contributions of Fatal Cancers of Different Sites to the Detriment Induced by Radiation

13.1 Introduction

The radiation protection system introduced by ICRP in 1977 (ICRP, 1977) and modified in 1991 (ICRP, 1991) attempted to quantify the late occurring stochastic effects of radiation after low doses of radiation and to apportion the risks incurred among organs and tissues by assigning weighting factors to them. This system was adopted by NCRP in 1987 (NCRP, 1987). Application of the system requires either that its weighting factors be applicable to many different circumstances, including populations of different age and sex, or that different sets of weights be used for these different circumstances.

The starting point for the evaluation of data pertaining to weighting factors is the relative probability of excess fatal cancer (lifetime) in the different organs and tissues. The UNSCEAR Report (UNSCEAR, 1988) provided such data; see Table 13.1 for organ and tissue risk coefficients based on the evaluation of the atomic-bomb survivor data by Shimizu *et al.* (1988). However, age-averaged tissue or organ-weighting factors insensitive to age variations were used. The BEIR V Report (NAS/NRC, 1990, Table 4-3) also provided some information on risks for a few organs and organ systems as a function of age for the United States population. Both of these sources are useful, but do not provide risk data in sufficient detail about the dependence of organ risks on variables such as age, sex, national characteristics and population-transfer model to be adequate for radiation protection purposes.

Land and Sinclair (1991) examined the dependence of relative probabilities of fatal cancer for those organs listed by UNSCEAR (1988), on sex, age, national population and population-transfer

TABLE 13.1—*Excess lifetime probability of a fatal cancer (specific) after acute whole-body exposure to 1 Gy organ absorbed dose of low-LET radiation (adapted from Table 69, Annex F, UNSCEAR, 1988).*[a,b]

Malignancy	Probability of fatal cancer (10^{-2})	
	Multiplicative risk Projection model	Additive risk Projection model
Red bone marrow	0.97	0.93
All cancers except leukemia	6.10	3.60
Bladder	0.39	0.23
Breast[c]	0.60	0.43
Colon	0.79	0.29
Lung	1.51	0.59
Multiple myeloma	0.22	0.09
Ovary[c]	0.31	0.26
Esophagus	0.34	0.16
Stomach	1.26	0.86
Remainder	1.14[d]	1.03[d]
	1.18[e]	0.66[e]
Total	7.07[f]	4.53[f]
	7.12[g]	4.16[g]

[a]Estimates based on Japanese population.

[b]Estimates based on age-averaged coefficients.

[c]These values must be divided by two to calculate the total and other organ probability values. Values are similar for Japanese survivors and other sources.

[d]This value is derived by subtracting the sum of the probabilities at the sites specified from the probabilities for all cancers except leukemia.

[e]This value is derived by fitting a linear relative probability model to the basic cancer data after the exclusion of those cases of cancer at the specific sites listed. (Coefficients: 0.19 excess relative probability per Gy and 1.87×10^{-4} per person-y Gy).

[f]Red bone marrow plus all other cancers.

[g]Red bone marrow plus other individual sites including remainder.

model. Age-specific data from the studies of the atomic-bomb survivors (Shimizu *et al.*, 1988) were used when available, and age-averaged coefficients were used for organs such as the esophagus, ovary and bladder for which age-specific data were not available. Two models; (1) the multiplicative model (UNSCEAR, 1988), which involves multiplicative transfer from one population to another and projection in time multiplicatively; and (2) the NIH-projection model (NIH, 1985), which involves additive transfer between populations and then multiplicative projection in time, were used. A third model, additive in both aspects (UNSCEAR, 1988), was also examined by Land and Sinclair (1991) but not considered appropriate. In Figure 13.1, the impact of the choice of projection model on the estimate of the risk is shown. Five national populations were considered: Japan, United States, United Kingdom, Puerto Rico and China; also considered were both sexes and four age ranges: 0 to 19 y, 20 to 64 y, 65 to 90 y and 0 to 90 years.

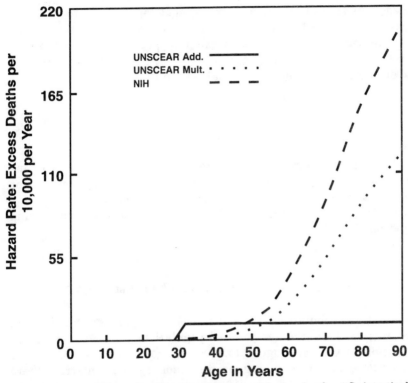

Fig. 13.1. Excess mortality from all cancers except leukemia after 1 Sv intestinal equivalent dose at age 20 using the UNSCEAR additive: ——, UNSCEAR multiplicative: ····· and NIH: — —, projection models to estimate excess mortality in United States males (Land and Sinclair, 1991).

The estimates for excess cancer mortality made by UNSCEAR (1988) did not include a DDREF. The BEIR V (NAS/NRC, 1990) used a linear-quadratic model for leukemia which resulted in estimates of risk for low doses and low-dose rates that were about a factor of two less than for high doses. ICRP used a reduction factor of two. NCRP considers that values of DDREF are probably dependent on the tissue involved and that for cancers in certain tissues, and for life shortening, the values of DDREF are likely to exceed two. However, as few data exist for humans on the relative influence of high and low doses on the risk of solid cancers, NCRP concurs with the selection of a factor of two made by ICRP (1991).

13.2 Sex

The differences in the relative contribution of the organs to the total fatal cancer risk for males and females is shown for the two models in Table 13.2 for the population of Japan, age 0 to 90 years. Evidently sex differences occur and while often small, they range up to about a factor of three, for example, in the case of the bladder. In the case of the multiplicative model, the risk for all cancers is about 30 percent greater in females than in males.

13.3 Age

The relative probabilities of fatal cancer in the different organs vary with age (Table 13.3), and the total risk is greater, by a factor of about three, in the 0 to 19 y group than in the 20 to 64 y group. The choice of model makes little difference to the total risk (Land and Sinclair, 1991).

13.4 National Populations and Transfer Models

Relative probabilities of fatal cancer in organs for the different characteristics of the five national populations considered using the multiplicative model, ages 0 to 90 y, are shown in Table 13.4. While the differences for many organs are minor, e.g., bone marrow, there are considerable differences for organs such as esophagus (a factor

TABLE 13.2—*Relative probabilities of fatal cancer in specific organs in relation to sex (United States population, age 0 to 90 y) (adapted from Table 5, Land and Sinclair, 1991).*

| Organ | Risk models[a] | | | |
| | Multiplicative | | NIH | |
	Male	Female	Male	Female
Esophagus	0.013	0.015	0.020	0.030
Stomach	0.039	0.028	0.279	0.354
Colon	0.269	0.369	0.224	0.152
Lung	0.266	0.143	0.129	0.112
Breast	—	0.149	—	0.067
Ovary	—	0.062	—	0.047
Bladder	0.122	0.029	0.070	0.026
Bone marrow	0.139	0.054	0.125	0.061
Residual	0.150	0.150	0.150	0.150
	Total probability $(10^{-2} \text{ Sv}^{-1})$			
	9.1	13.3	7.7	9.6

[a]In both the multiplicative and NIH models a constant relative risk over time after exposure was assumed. The transfer of the risk across populations was multiplicative in the multiplicative risk model and additive in the NIH model (NIH, 1985).

TABLE 13.3—*Relative probabilities of fatal cancer in organs in two age groups at exposure based on the average for male and female in the United States population (adapted from Table 5, Land and Sinclair, 1991).[a]*

Organ	Risk models			
	Multiplicative		NIH	
	0–19 y	20–64 y	0–19 y	20–64 y
Esophagus	0.008	0.024	0.008	0.048
Stomach	0.030	0.040	0.300	0.347
Colon	0.423	0.174	0.272	0.066
Lung	0.223	0.187	0.130	0.105
Breast	0.074	0.083	0.043	0.022
Ovary	0.017	0.057	0.021	0.027
Bladder	0.040	0.125	0.028	0.073
Bone marrow (leukemia)	0.034	0.160	0.048	0.160
Residual	0.150	0.150	0.150	0.150
Total probability (10^{-2} Sv^{-1})				
	23.5	8.3	18.2	6.5

[a]The probabilities of fatal cancer in United States population were obtained by transfer between the LSS of the atomic-bomb survivors (1950 to 1985) to a United States population that reflects current cancer rates for 1973 to 1977.

TABLE 13.4—*Relative probabilities of fatal cancer in organs for five populations. Average of male and female, age 0 to 90 y, multiplicative model[a] (from Table 5, Land and Sinclair, 1991).[b]*

Organ	Japan	United States	Puerto Rico	United Kingdom	China	Average
Esophagus	0.038	0.014	0.098	0.030	0.269	0.090
Stomach	0.291	0.033	0.136	0.050	0.224	0.146
Colon	0.180	0.320	0.206	0.225	0.103	0.207
Lung	0.174	0.205	0.141	0.274	0.097	0.178
Breast	0.023	0.075	0.048	0.085	0.022	0.051
Ovary	0.014	0.031	0.016	0.031	0.019	0.022
Bladder	0.052	0.076	0.078	0.090	0.036	0.066
Bone marrow	0.077	0.096	0.127	0.064	0.079	0.089
Remainder	0.150	0.150	0.150	0.150	0.150	0.150
All cancer	0.999	1.000	1.000	0.999	0.999	0.999
Total probability (10^{-2} Sv^{-1})						
	10.2	11.2	9.5	12.9	6.4	10.0

[a]The multiplicative model was used for both the transfer of risk across populations and the projection across time.

[b]The values in the table have been calculated directly from Table 5 of Land and Sinclair (1991). No attempt has been made to force the relative probabilities for all cancers to the logical values of one.

of up to 19) and stomach (a factor of up to nine). Using the NIH model, the variation is no more than about a factor of two for any organ (see Table 13.5). This does not, however, imply that the NIH model is superior since there is no general agreement, at the present

TABLE 13.5—*Relative probabilities of fatal cancer in organs for five populations. Average for male and female, age 0 to 90 y, NIH projection model[a] (derived from Table 5, Land and Sinclair, 1991).[b]*

Organ	Japan	United States	Puerto Rico	United Kingdom	China	Average
Esophagus	0.042	0.025	0.030	0.023	0.037	0.031
Stomach	0.268	0.316	0.345	0.336	0.291	0.311
Colon	0.121	0.188	0.138	0.146	0.113	0.141
Lung	0.220	0.120	0.137	0.182	0.132	0.158
Breast	0.027	0.034	0.027	0.028	0.044	0.032
Ovary	0.019	0.023	0.027	0.019	0.022	0.022
Bladder	0.052	0.048	0.054	0.037	0.052	0.049
Bone marrow	0.100	0.093	0.090	0.076	0.158	0.103
Remainder	0.150	0.150	0.150	0.150	0.150	0.150
All cancer	0.999	0.997	0.998	0.997	0.999	0.997
Total probability (10^{-2} Sv^{-1})						
	9.3	8.7	10.1	9.7	6.0	8.8

[a]The absolute risk was used for the transfer across populations and the multiplicative model for projection across time.

[b]The values in the table have been calculated directly from Table 5 of Land and Sinclair (1991). No attempt has been made to force the relative probabilities for all cancers to the logical values of one.

time, on which transfer model to use in given sites or even whether it should be the same for all sites. Testing of models would be possible if independent sources of induced cancer information were available in one of the populations to which a transfer has been made. This is the case with cancer of the breast in the United States population, but unfortunately sources of information here have been used to support both the NIH model (NIH, 1985) and to support the multiplicative model (NAS/NRC, 1990).[9]

13.5 United States Population Versus the International Commission on Radiological Protection "Average Population"

The International Commission on Radiological Protection (ICRP, 1991) decided, in view of the differences between transfer models and populations, to average over five populations and over sex, for both models, and to specify a single set of relative probabilities of fatal cancer in the various organs. The largest difference between

[9]For breast cancer incidence, the data are now definite in that the NIH model (NIH, 1985) is strongly supported and the multiplicative model should be rejected (Preston, 1993).

the ICRP average and the United States is for the esophagus, which represents a relatively small contribution to the total. The differences for all other organs are relatively small (see Table 13.6). Furthermore, the risk of cancer mortality for the United States population, 10×10^{-2} Sv^{-1}, and the average for the five populations of 9.5×10^{-2} Sv^{-1} as determined by ICRP (1991) are in good agreement.

13.6 Expected Years of Life Lost from Fatal Cancer in Organs Versus Sex, Age, Population and Model

In the same way as for the probability of fatal cancer, the expected years of life lost for each fatal cancer can be tabulated and averaged (see Land and Sinclair, 1991, for details). These yield a table (see Table 13.7) of relative probabilities in organs that are quite similar to those for fatal cancer probabilities except that the contribution of bone marrow is much greater by about a factor of two. This is due to the relatively short latent period for leukemia.

13.7 Comparison of Estimates of Fatal Cancer Risks in Organs in 1991 With Those of 1977

The estimates of the risk of fatal cancer (lifetime) in organs for the whole population for ICRP (1991) and this Report are given in

TABLE 13.6—*Distribution of probabilities of fatal cancer in organs.*

Organ	ICRP average[a]	U.S.[b]
Esophagus	0.061	0.020
Stomach	0.229	0.175
Colon	0.174	0.254
Lung	0.168	0.163
Breast	0.041	0.055
Ovary	0.022	0.027
Bladder	0.058	0.062
Bone marrow	0.096	0.094
Remainder	0.150	0.150
All cancer	0.999	1.000
Total probability (10^{-2} Sv^{-1})	9.5	10.0

[a]Average of males and females, five national populations, two transfer models (multiplicative and NIH), age 0 to 90 years (from Table B-15, ICRP, 1991).

[b]Average of males and females, two transfer models (multiplicative and NIH), age 0 to 90 years (calculated from Table 5, Land and Sinclair, 1991).

TABLE 13.7—*Years of life lost and relative values of expected life lost due to induced cancer among organs averaged for five national populations and two models (multiplicative and NIH), age 0 to 90 y (adapted from Tables B-16 and B-18, ICRP, 1991).*

Organ	Life lost (years)	Relative years of life lost
Esophagus	11.5	0.048
Stomach	12.4	0.190
Colon	12.5	0.148
Lung	13.5	0.154
Breast	18.2	0.049
Ovary	16.8	0.025
Bladder	9.8	0.039
Bone marrow	30.9	0.197
Remainder	13.7	0.150
All cancer	15.0[a]	1.000

[a]Calculated by dividing expected years of life lost for all cancer by the total number of fatal cancers.

Table 13.8. These are compared with estimates made in 1977 (ICRP, 1977). The overall estimate of risk is higher, but for some of the organs the risks are about the same, *e.g.*, breast, bone, thyroid. Others are increased, such as lung and bone marrow, while still

TABLE 13.8—*Lifetime mortality in a population of all ages from specific fatal cancer after exposure to low doses (i.e., DREF of two applied) (from Table B-17, ICRP, 1991).*

	Fatal probability coefficient 10^{-4} Sv^{-1}	
	ICRP (1977) and NCRP (1987)[a]	ICRP (1991) and this Report
Bladder	—	30
Bone marrow	20	50
Bone surface	5	5
Breast	25	20
Colon	—	85
Liver	—	15
Lung	20	85
Esophagus	—	30
Ovary	—	10
Skin	—	2
Stomach	—	110
Thyroid	5	8
Remainder[b]	50	50
Total	125[c]	500[d]

[a]Calculated from tissue weighting factors and a total risk of 165×10^{-4} Sv^{-1}.
[b]The composition of the remainder is quite different in the two cases.
[c]This total was used for both the worker population and the general public.
[d]General public only. The total fatal cancer risk for a working population is taken to be 400×10^{-4} Sv^{-1}.

others, for the first time, make a large contribution to the total risk. Cancers of the colon and stomach were not included separately in the risk estimates in 1977 but were part of the remainder. They now contribute almost 40 percent of the total risk. (Note that these estimates are for exposure to low doses and have been derived from estimates from high-dose-rate exposure using a DREF of two.)

14. Detriment Due to Radiation Exposure at Low Doses

14.1 Introduction

The estimate of the total detriment that results from exposure to radiation is complex and is still the subject of debate (ICRP, 1985; 1991). Four major components of the detriment are considered: (1) the risk of fatal cancer, (2) the risk of severe hereditary disease, (3) the relative loss of lifespan for different cancers and for hereditary disease, and (4) a contribution from nonfatal cancer. ICRP (1991) did not include any possible detriment arising from mental retardation in the fetus and none is included in this Report. If a threshold for mental retardation is justified, then no detriment need be considered for the range of doses applicable to routine radiation protection. If not, a calculation and allowance can be made, for example, Sinclair (1992). However, in view of the uncertainty about a threshold, this possible detriment is not included as a component of the detriment.

14.2 Estimate of the Detriment

The first component, the probability of fatal cancer for each organ, is obtained from the data shown in Table 13.8. The total probability of fatal cancer in the general population is 5×10^{-2} Sv^{-1} and 4×10^{-2} Sv^{-1} in a working population. The contribution of severe hereditary disease is about 1.3×10^{-2} Sv^{-1} for members of the general population and 0.8×10^{-2} Sv^{-1} for a working population. The third component, the relative length of life lost due to radiation-induced cancer in specific organs, is obtained from the data in Table 13.7. The relative years of life lost varies by a factor of about seven to eight among the fatal cancers listed.

The assessment of the contribution of the last component of the total detriment, nonfatal cancer, is based on the severity or lethality of the specific cancer and weighted according to the lethality fraction, k

(ICRP, 1991). In calculating total detriment for a particular type of cancer, fatal cases receive full weight and a partial weight, p, is placed on nonfatal cases. Thus, if T is the total number of cancer cases and k the fatality fraction, the detriment (D) is:

$$D = kT + p(1 - k)T. \tag{14.1}$$

The ICRP somewhat arbitrarily chose $p = k$, reasoning that a nonfatal cancer of a type that is usually fatal involves more unpleasantness to the patient than a nonfatal cancer of a type that is seldom fatal. With this choice, D is:

$$D = kT + k(1 - k)T = kT(2 - k) = F(2 - k) \tag{14.2}$$

where $F = kT$ is the number of fatal cancers.

Two sets of data have been used to select the lethality fractions: (1) the 5 y survival rates for cancer of the specific organs for the years 1980 to 1985 and (2) the lethality rates for cancer based on the years 1950 to 1970. Neither set of data alone is appropriate, the 5 y rates underestimate the lethality and the 1950 to 1970 rates overestimate the lethality because of the improvement in cancer-cure rates in recent years. Therefore, judgment, as well as the data available for 5 and 20 y lethality rates, have been used in the selection of the lethality fractions shown in Table 14.1.

TABLE 14.1—*Lethality data for cancers in adults by site (NIH, 1989)*[a] *[from Table B-19, ICRP Publication 60 (ICRP, 1991)].*

	Five years 1980 to 1985	Twenty years lethality 1950 to 1970	Proposed lethality Fraction (k)
Bladder	0.22	0.58	0.50
Bone	—	0.72	0.70
Brain	0.75	0.84	0.80
Breast	0.24	0.62	0.50
Cervix	0.33	0.50	0.45
Colon	0.45	0.62	0.55
Kidney	0.48	0.78	0.65
Leukemia (acute)	0.98	0.99	0.99
Liver	0.95	0.98	0.95
Lung and bronchus	0.87	0.96	0.95
Esophagus	0.92	0.97	0.95
Ovary	0.62	0.74	0.70
Pancreas	0.97	0.99	0.99
Prostate	0.26	0.84	0.55
Skin	—	—	0.002
Stomach	0.85	0.90	0.90
Thyroid	0.06	0.15	0.10
Uterus	0.17	0.35	0.30

[a]Numbers were derived from tables and graphical data in NIH (1989).

The expected years of life lost as a result of radiation-induced cancer differs among organs and the differences must be taken into account in the estimate of total detriment. Land and Sinclair (1991) estimated the years of life lost for each cancer (l) and the average years of life lost ($\bar{l}$) for all cancers which was 15 years. The relative expected life lost ($l/\bar{l}$) (lethality factor, k) is based on an average for sex, age at exposure, national population and both the multiplicative and NIH models.[10] The length of life lost for cancer of breast and ovary are based only on the data for females. The years of life lost for severe genetic effects was taken to be an average of 20 y and, therefore, a correction factor for genetic effects of 20 y divided by 15 y or 1.33. The k for each cancer shown in Tables 14.1 and 14.2, are derived from $l/\bar{l}$ where l is the expected years of life lost for a specific site and $\bar{l}$ is the average for all cancers.

14.3 Tissue Weighting Factors

The relative contribution of the individual organs to the total detriment is shown in Table 14.2 and form the basis of the weighting factors to be used. There are large uncertainties in the data at each stage of the derivation of these organ contributions. Consequently, it is reasonable to round off the values for the various organs (Table 14.2) and limit the weighting factors to four groups. The ICRP in its Publication 60 (ICRP, 1991) concluded that a reasonable assignment of weights derived from the data of Table 14.2 for the four groups was that which is shown in Table 14.3. NCRP recommends the same weighting factors since the United States population does not differ in important respects from the ICRP average population. The dependence of the weighting factors on the various variables has been assessed.

14.4 Working Populations

Although there are considerable uncertainties in the estimates of the four components that contribute to the total detriment, it is reasonable to examine whether tissue weighting factors should be different for the working population from those for the general population. For example, the nominal probability of radiation-induced

---

[10]These symbols, l and $\bar{l}$, are from ICRP (1991).

TABLE 14.2—Relative contribution of organs to the total detriment for a general population [adapted from Table B-20, ICRP Publication 60 (ICRP, 1991)].

	Probability of fatal Cancer (F) (per 10,000 person-Sv)	Severe Genetic Effects (per 10,000 person-Sv)	Relative Length of life Lost $l/\bar{l}$	Relative Nonfatal Contribution $(2-k)$	Product of $F[l/\bar{l}(2-k)]$ (per 10,000 person-Sv)	Relative Contribution	Rounded Weight (w_T)
Bladder	30		0.65	1.50	29.4	0.040	0.05
Bone marrow	50		2.06	1.01	104.0	0.143	0.12
Bone surface	5		1.00	1.30	6.5	0.009	0.01
Breast	20		1.21	1.50	36.4	0.050	0.05
Colon	85		0.83	1.45	102.7	0.141	0.12
Liver	15		1.00	1.05	15.8	0.022	0.05
Lung	85		0.90	1.05	80.3	0.111	0.12
Esophagus	30		0.77	1.05	24.2	0.034	0.05
Ovary	10		1.12	1.30	14.6	0.020	0.05
Skin	2		1.00	2.00	4.0	0.006	0.01
Stomach	110		0.83	1.10	100.0	0.139	0.12
Thyroid	8		1.00	1.90	15.2	0.021	0.05
Remainder	50		0.91	1.29	58.9	0.081	0.05
Gonads		100	1.33	—	133.3	0.183	0.20
Total	500		—	—	725.3	1.000	—

TABLE 14.3—*Tissue weighting factors, w_T (ICRP, 1991).*

w_T		Σw_T
0.01	bone surface, skin	0.02
0.05	bladder, breast, liver, esophagus, thyroid, remainder	0.30
0.12	bone marrow, colon, lung, stomach	0.48
0.20	gonads	0.20
	Total	1.00

fatal cancer in a lifetime is 4×10^{-2} Sv^{-1} for a working population which is about 80 percent of the total cancer risk for the general population. The total risk is lower because of the absence of younger persons who have a higher susceptibility for radiation-induced cancer. However, the differences in the total detriment between the general and working populations are relatively small. In addition, when the individual contributions to the total detriment are rounded off and assigned to only four groups, the same weighting factors appear to be appropriate for both populations. Therefore, the weighting factors shown in Table 14.3 should be used for both a working population and the general population.

14.5 Total Detriment

The total detriment after low-dose, low-dose-rate exposure to ionizing radiation is the sum of the contributions due to fatal cancer and severe hereditary disorders weighted for length of life lost. These are listed in Table 14.4 for a working population and for the whole population. The total detriment attributed to stochastic effects is 7.3×10^{-2} Sv^{-1} for the whole population and some 23 percent less for a working population of 20 to 64 y of age.

The data in Tables 14.3 and 14.4 provide the currently recommended nominal risks and organ and tissue weights to be applied in setting radiation protection standards.

TABLE 14.4—*Nominal detriment coefficients for stochastic effects (10^{-2} Sv^{-1})[a] (from Table S-3, ICRP, 1991).*

Exposed Population	Fatal Cancer[b]	Nonfatal Cancer	Hereditary Disorders	Total
Working population 20 to 64 y of age	4.0	0.8	0.8	5.6
Whole population 0 to 90 y of age	5.0	1.0	1.3	7.3

[a]Rounded values.

[b]For fatal cancer, the detriment coefficient is the same as the probability coefficient.

15. Conclusions

It is clear that there are important unanswered questions about the estimation of risk of radiation-induced cancer. The NCRP has been particularly concerned about certain aspects of the estimates of the risk of radiogenic cancer. These include:

1. The choice of the most appropriate-risk projection model.
2. The duration of excess risk that is assumed in the risk-projection model. (For those persons exposed at a young age this has an important impact on the estimate of the lifetime risk.)
3. The estimates of the risks for persons exposed to low total doses over a considerable fraction of their life span. (It is these risks that are the concern for radiation protection of workers and the general population. In addition, these risks have had to be derived from data for persons exposed to high doses at high-dose rates.)
4. The fact that the estimates of risk of fatal cancer by the UNSCEAR (1988), BEIR V (NAS/NRC, 1990) and ICRP (1991) Committees are based on those for a Japanese population exposed in wartime and a reference Japanese population in 1982 rather than a current United States population. There are significant differences between the Japanese population and the United States population and the correct method of transferring radiation risk estimates across populations is not known.

Despite these nontrivial concerns, the NCRP considers that the data support an increase in the estimates of risks of radiogenic cancer from those made by UNSCEAR (1977) and BEIR III (NAS/NRC, 1980). The selection of a multiplicative risk model instead of an additive risk model has contributed significantly to the increases in the estimates of risk and the NCRP concurs with this choice of risk model.

The risk estimates reported by UNSCEAR (1988) and BEIR V (NAS/NRC, 1990) indicate a nominal value of 8×10^{-2} Sv^{-1} effective dose for a lifetime risk of cancer mortality in a worker population and 10×10^{-2} Sv^{-1} for the general population. These estimates are based on data obtained from persons exposed at high doses and high-dose rates. The NCRP considers that a DREF should be applied to

111

these estimates of risk for persons exposed to low doses and at low-dose rates and the factor could range between two and three. The effect of the value of the DREF on the risk coefficients for cancer mortality are shown in Table 15.1. The lifetime risk of fatal cancer as a result of exposure to low doses and at low-dose rates would, therefore, range between 2.7 to 4×10^{-2} Sv^{-1} for a worker population and 3.3 to 5×10^{-2} Sv^{-1} for the general population (see Table 15.1). **The NCRP, therefore, recommends that for radiation protection purposes, a DREF of two be applied and a lifetime risk of fatal cancer of 4×10^{-2} Sv^{-1} be used for a worker population and similarly, a lifetime risk of 5×10^{-2} Sv^{-1} be used for the general population.**

Other detrimental effects such as genetic effects, damage to the embryo/fetus and length of life lost must be considered in the assessment of the total detriment. The most recent estimates based on the limited human data available suggest that the human is not at as great a risk for genetic effects as was previously thought. On the other hand, the NCRP believes that the estimates based on experimental data may be higher than other committees have judged them to be (see Section 11). Perhaps the greatest uncertainty lies in the risk of multifactorial diseases. Given these uncertainties, the NCRP has chosen to recommend for radiation protection purposes a risk estimate of about 1.0×10^{-2} Sv^{-1} for severe hereditary effects in the total population. The risk estimate for a worker population is considered to be slightly lower.

The exposures of the embryo/fetus to radiation poses questions regarding both the estimates of the risk of cancer mortality, and also of deterministic effects. The systematic examination in recent years of the effects of radiation on the brain has clarified which is the most radiosensitive period in development, and has quantified the effects on intelligence and the probability of severe mental retardation. The important question of whether there is a threshold for the effects on brain has not been answered unequivocally. The potential risk that can result from exposure at the critical period of development, for example, between 8 to 15 weeks gestational age, necessitates strict limitation of exposure. The NCRP recommends that for

TABLE 15.1—*Risk coefficients for cancer mortality (average of UNSCEAR and BEIR V estimates) adjusted by varied values of DREF.*

	DREF		
	2	2.5	3
Population	Lifetime cancer mortality (10^{-2} Sv^{-1})		
Adult workers	4.0	3.2	2.7
Total population	5.0	4.0	3.3

radiation protection purposes, an estimate of 0.40×10^{-2} Sv^{-1} for severe mental retardation after exposure during the 8 to 15 week period of gestation be used.

Despite all the uncertainties, and the assumptions that must be made, the best estimates of risk of detrimental effects from exposure to radiation are higher than previously suggested and by a factor of about three for a working population. A summary of the estimated probabilities of the detrimental effects of low-LET radiation is shown in Table 14.2.

16.   Future Studies

It must have become clear to the reader of this Report that improvements in the methods of risk estimation, and the understanding of the underlying processes, are still required to improve the estimates of risk of radiation-induced health effects. The study of the atomic-bomb survivors over approximately 40 y have provided the data on which radiation protection standards have been set, and yet the data up to 1985 consist of about 340 cases of cancer out of approximately 6,000 cancer deaths that are attributed to radiation exposure. This is not a large number when all the necessary stratifications are made. Furthermore, it is not a large number in relation to the total number of persons still at risk since about 60 percent of the population exposed in the two cities is still alive. An increasing fraction of the surviving population is made up of those that were young ATB. This has entailed the use of methods of projecting the risk for those still alive to obtain an estimate of their lifetime risk, but it is not known over what period after exposure the excess risk exists or at what level. There is a need to understand how risk is related to age at exposure and how that risk is expressed with time. The approaches used in the BEIR V Report (NAS/NRC, 1990) for modeling the risk with time was a start, but more needs to be done. There are other aspects of the analysis of the data that require examination. For example, the independence of diseases is assumed in all analyses but the validity of this assumption has not been demonstrated.

The distribution of the types of cancer that occur naturally is quite different in the Japanese population from those of European or United States populations. It is a matter of concern that such a large fraction of the total risk is attributed to cancers of the digestive system. This suggests two areas that require study. First, an examination of how to transfer risk estimates from one population to another. The fact that ICRP (1991), based on the work of Land and Sinclair (1991), considered that the best estimates were to be obtained by averaging over five populations and two models of risk projection is sufficient evidence that more work is required. Second, studies of the risk of cancer of the various sites in the gastrointestinal tract in populations other than the Japanese are needed.

A major advantage of the study of the atomic-bomb survivors is that there were relatively large numbers of persons exposed to a broad range of doses and, therefore, dose-response relationships can be assessed. However, the most important information for radiation protection purposes is the risk estimates for the affects of exposure to low doses and to protracted exposures. The atomic-bomb survivors were exposed to irradiation at a very high-dose rate and the data does not allow direct estimates of risk at low doses or low-dose rates. Further studies are needed that will provide useful determinations of the effect of dose rate and fractionation. There is no comprehensive model of time-dose relationships.

New studies are revealing more and more about the carcinogenic process at a molecular level. It is becoming accepted that carcinogenesis is a multistage process. It is not known whether irradiation effects steps in the carcinogenic process beyond the initial events and how dose rate affects the mutational events considered important for cancer induction. It is the factors influencing expression of the initial events that are the determinants of whether a cancer actually develops. Little is known about how radiation influences these factors. There is still the hope that radiation-specific lesions in DNA will be identified. Such identification would make it possible to establish that a specific cancer was or was not attributable to the radiation exposure irrespective of dose.

In the field of genetics, some fundamental questions remain to be answered. For example, one of the sources of uncertainty in the application of the results obtained by the "doubling dose" method, which is used for estimating the genetic risk, is whether the human immature oocyte is more or less radiosensitive than the spermatogonia. The number of genes, the spontaneous mutation rates of a number of critical genes involved in major genetic diseases and the precise nature of the alteration induced in the genome by radiation, are all questions to be answered. Perhaps the most important question to answer is about the importance of the category of genetic disorders called multifactorial disorders involving the interaction of more than one gene. A critical question is what is the magnitude of the mutational component of these diseases? The effect of genetic diseases on life span is poorly understood and, therefore, the impact of a radiation-induced excess of these diseases has not been adequately assessed.

New techniques make it possible to determine, in humans, induced aneuploidy and chromosome rearrangements with defined exposure. An example of what should be done is the determination of the relative frequencies of aneuploidy for chromosomes 13, 18, 21 and the six chromosomes which comprise the main class of chromosome diseases in the live born.

Lastly, studies with the potential to establish incontrovertible evidence that a threshold for radiation-induced mental retardation, or reduction in the IQ does or does not exist, are of great importance. It is imperative that further studies on the mechanisms of radiation-induced effects of the brain be carried out.

The required studies are at all levels of biological organization, and the skills needed range from epidemiology to molecular biology.

References

ABRAHAMSON, S. (1990a). "Childhood leukemia at Sellafield," Radiat. Res. **123**, 237–238.

ABRAHAMSON, S. (1990b). "Genetic risk estimates. Of mice and men," pages 61 to 69 in *Radiation Protection Today—The NCRP at Sixty Years*, NCRP Proceedings No. 11 (National Council on Radiation Protection and Measurements, Bethesda, Maryland).

ABRAHAMSON, S. (1990c). "Risk estimates: Past, present, and future," Health Phys. **59**, 99–102.

BEEBE, G.W., KATO, H. and LAND, C.E. (1978a). "Studies of the mortality of A-bomb survivors. 6. Mortality and radiation dose, 1950–1974," Radiat. Res. **75**, 138–201.

BEEBE, G.W., LAND, C.E. and KATO, H. (1978b). "The hypothesis of radiation-accelerated aging and the mortality of Japanese A-bomb victims," pages 3 to 27 in *Late Biological Effects of Ionizing Radiation*, Volume 1, STI/PUB/489 (International Atomic Energy Agency, Vienna).

BERAL, V., FRASER, P., CARPENTER, L., BOOTH, M., BROWN, A. and ROSE, G. (1988). "Mortality of employees of the Atomic Weapons Establishment, 1951–82," Br. Med. J. **297**, 757–770.

BITHELL, J.F. and STILLER, C.A. (1988). "A new calculation of the carcinogenic risk of obstetric x-raying," Stat. Med. **7**, 857–864.

BLOT, W.J. and MILLER, R.W. (1973). "Mental retardation following *in utero* exposure to the atomic bombs of Hiroshima and Nagasaki," Radiology **106**, 617–619.

BOICE, J.D., JR., DAY, N.E., ANDERSEN, A., BRINTON, L.A., BROWN, R., CHOI, N.W., CLARKE, E.A., COLEMAN, M.P., CURTIS, R.E., FLANNERY, J.T., HAKAMA, M., HAKULINEN, T., HOWE, G.R., JENSEN, O.M., KLEINERMAN, R.A., MAGNIN, D., MAGNUS, K., MAKELA, K., MALKER, B., MILLER, A.B., NELSON, N., PATTERSON, C.C., PETTERSSON, F., POMPE-KIRN, V., PRIMIC-ŽAKELJ, M., PRIOR, P., RAVNIHAR, B., SKEET, R.G., SKJERVEN, J.E., SMITH, P.G., SOK, M., SPENGLER, R.F., STORM, H.H., STOVALL, M., TOMKINS, G.W.O. and WALL, C. (1985). "Second cancers following radiation treatment for cervical cancer. An international collaboration among cancer registries," J. Natl. Cancer Inst. **74**, 955–975.

BOICE, J.D., JR., BLETTNER, M., KLEINERMAN, R.A., STOVALL, M., MOLONEY, W.C., ENGHOLM, G., AUSTIN, D.F., BOSCH, A., COOKFAIR, D.L., KREMENTZ, E.T., LATOURETTE, H.B., PETERS, L.J., SCHULZ, M.D., LUNDELL, M., PATTERSSON, F., STORM, H.H., BELL, C.M.J., COLEMAN, M.P., FRASER, P., PALMER, M., PRIOR, P., CHOI, N.W., HISLOP, T.G., KOCH, M., ROBB, D., ROBSON, D., SPRENGLER, R.F., VON FOURNIER, D., FRISCHKORN, R.,

117

LOCHMÜLLER, H., POMPE-KIRN, V., RIMPELS, A., KJORSTAD, K.E., PEJOVIC, M.H., SIGURDSSON, K., PISANI, P., KUCERA, H. and HUTCHISON, G.B. (1987). "Radiation dose and leukemia risk in patients treated for cancer of the cervix," J. Nat. Cancer Inst. **79**, 1295–1311.

BOICE, J.D., JR., ENGHOLM, G., KLEINERMAN, R.A., BLETTNER, M., STOVALL, M., LISCO, H., MOLONEY, W.C., AUSTIN, D.F., BOSCH, A., COOKFAIR, D.L., KREMENTZ, E.T., LATOURETTE, H.B., MERRILL, J.A., PETERS, L.J., SCHULZ, M.D., STORM, H.H., BJÖRKHOLM, E., PETTERSSON, F., BELL, C.M.J., COLEMAN, M.P., FRASER, P., NEAL, F.E., PRIOR, P., CHOI, N.W., HISLOP, T.G., KOCH, M., KREIGER, N., ROBB, D., ROBSON, D., THOMSON, D.H., LOCHMÜLLER, H., VON FOURNIER, D., FRISCHKORN, R., KJORSTAD, K.E., RIMPELA, A., PEJOVIC, M.H., POMPE-KIRN, V., STANKUSOVA, H., BERRINO, F., SIGURDSSON, K., HUTCHISON, G.B. and MACMAHON, B. (1988). "Radiation dose and second cancer risk in patients treated for cancer of the cervix," Radiat. Res. **116**, 3–55.

CALDWELL, G.G., KELLEY, D., ZACK, M., FALK, H. and HEATH, C.W., JR. (1983). "Mortality and cancer frequency among military nuclear test (Smoky) participants, 1957 through 1979," JAMA **250**, 620–624.

CARDIS, E. and KALDOR, J. (1989). *Protocol of Combined Analysis of Cancer Mortality Among Nuclear Workers*, Internal Report No. 89/005 (International Agency for Research on Cancer, Lyon, France).

CHMELEVSKY, D., KELLERER, A.M., SPIESS, H. and MAYS, C.W. (1986). "A proportional hazards analysis of bone sarcoma rates in German ^{224}radium patients," pages 32 to 37 in *The Radiobiology of Radium and Thorotrast*, Gössner, W., Gerber, G.B., Hagen, U. and Luz, A., Eds. (Urban and Schwarzenberg, Baltimore, Maryland).

COX, D.R. (1962). *Renewal Theory* (Wiley, New York).

CZEIZEL, A., SANKARANARAYANAN, K., LOSONCI, A., RUDAS, T. and KERESZTES, M. (1988). "The load of genetic and partially genetic diseases in man. II. Some selected common multifactorial diseases: Estimates of population prevalence and of detriment in terms of years of lost and impaired life," Mutat. Res. **196**, 259–292.

DARBY, S.C. and DOLL, R. (1987). "Fallout, radiation doses near Dounreay, and childhood leukaemia," Br. Med. J. **294**, 603–607.

DARBY, S.C., NAKASHIMA, E. and KATO, H. (1985). "A parallel analysis of cancer mortality among atomic bomb survivors and patients with ankylosing spondylitis given x-ray therapy," J. Natl. Cancer Inst. **75**, 1–21.

DARBY, S.C., DOLL, R., GILL, S.K. and SMITH, P.G. (1987). "Long term mortality after a single treatment course with X-rays in patients treated for ankylosing spondylitis," Brit. J. Cancer **55**, 179–190.

DARBY, S.C., KENDALL, G.M., TELL, T.P. and O'HAGAN, J.A. (1988). "A summary of mortality and incidence of cancer in men from the United Kingdom who participated in the United Kingdom's atmospheric nuclear weapons tests and experimental programmes," Br. Med. J. **296**, 332–338.

DARBY, S.C., MUIRHEAD, C.R., DOLL, R., KENDALL, G.M. and THAKRAR, B. (1990). "Mortality among United Kingdom servicemen who served abroad in the 1950s and 1960s," Br. J. Ind. Med. **47**, 793–804.

GARDNER, M.J., SNEE, M.P., HALL, A.J., POWELL, C.A., DOWNES, S. and TERRELL, J.D. (1990). "Results of case-control study of leukaemia and lymphoma among young people near Sellafield nuclear plant in West Cumbria," Brit. Med. J. **300**, 423–429.

GILBERT, E.S. and MARKS, S. (1979). "An analysis of the mortality of workers in a nuclear facility," Radiat. Res. **79**, 122–148.

GILBERT, E.S. and OHARA, J.L. (1984). "An analysis of various aspects of atomic bomb dose estimation at RERF using data on acute radiation symptoms," Radiat. Res. **100**, 124–138.

GILBERT, E.S., FRY, S.A., WIGGS, L.D., VOELZ, G.L., CRAGLE, D.L. and PETERSEN, G.R. (1989). "Analyses of combined mortality data on workers at the Hanford Site, Oak Ridge National Laboratory, and Rocky Flats Nuclear Weapons Plant," Radiat. Res. **120**, 19–35.

GRIFFIN, C.S. and TEASE, C. (1988). "Gamma-ray-induced numerical and structural chromosome anomalies in mouse immature oocytes," Mutat. Res. **202**, 209–213.

HARVEY, E.B., BOICE, J.D., JR., HONEYMAN, M. and FLANNERY, J.T. (1985). "Prenatal x-ray exposure and childhood cancer in twins," N. Engl. J. Med. **312**, 541–545.

HECKMAN, J.J. and HONORE, B.E. (1989). "The identifiability of the competing risks model," Biometrika **76**, 325–330.

HOLM, L.E., WIKLUND, K.E., LUNDELL, G.E., BERGMAN, N.A., BJELKENGREN, G., CEDERQUIST, E.S., ERICSSON, U.B., LARSSON, L.G., LIDBERG, M.E., LINDBERG, R.S., WICKLUND, H.V. and BOICE J.D., JR. (1988). "Thyroid cancer after diagnostic doses of iodine-131: A retrospective cohort study," J. Natl. Cancer Inst. **80**, 1132–1138.

HOLM, L.E., WIKLUND, K.E., LUNDELL, G.E., BERGMAN, N.A., BJELKENGREN, G., ERICSSON, U.B., CEDERQUIST, E.S., LIDBERG, M.E., LINDBERG, R.S., WICKLUND, H.V. and BOICE, J.D., JR. (1989). "Cancer risk in population examined with diagnostic doses of ^{131}I," J. Natl. Cancer Inst. **81**, 302–306.

HOWE, G.R., WEEKS, J.L., MILLER, A.B., CHIARELLI, A.M. and ETEZADI-AMOLI, J. (1987). *A Study of the Health of the Employees of Atomic Energy of Canada Limited. IV. Analysis of Mortality During the Period 1950–1981*, AECL-9442 (Atomic Energy of Canada Limited, Pinawa, Manitoba).

HRUBEC, Z., BOICE, J.D., JR., MONSON, R.R. and ROSENSTEIN, M. (1989). "Breast cancer after multiple chest fluoroscopies: Second follow-up of Massachusetts women with tuberculosis," Cancer Res. **49**, 229–234.

ICRP (1977). International Commission on Radiological Protection. *Recommendations of the International Commission on Radiological Protection*, ICRP Publication 26, Annals of the ICRP **1** (Pergamon Press, Elmsford, New York).

ICRP (1985). International Commission on Radiological Protection. *Quantitative Bases for Developing a Unified Index of Harm*, ICRP Publication 45, Annals of the ICRP **15** (Pergamon Press, Elmsford, New York).

ICRP (1987). International Commission on Radiological Protection. *Lung Cancer Risk from Indoor Exposures to Radon Daughters*, ICRP Publication 50, Annals of the ICRP **17** (Pergamon Press, Elmsford, New York).

ICRP (1991). International Commission on Radiological Protection. *1990 Recommendations of the International Commission on Radiological Protection*, ICRP Publication 60, Annals of the ICRP **21** (Pergamon Press, Elmsford, New York).

ICRP (1992). International Commission on Radiological Protection. *The Biological Basis for Dose Limitation in the Skin*, ICRP Publication 59, Annals of the ICRP **22** (Pergamon Press, Elmsford, New York).

JABLON, S. (1971). *Atomic Bomb Radiation Dose Estimation at ABCC*, Atomic Bomb Casualty Commission Technical Report 23-71 (Radiation Effects Research Foundation, Hiroshima).

JABLON, S., HRUBEC, Z. and BOICE, J.D., JR. (1991). "Cancer in populations living near nuclear facilities. A survey of mortality nationwide and incidence in two states," JAMA **265**, 1403–1408.

KATO, H. and SCHULL, W.J. (1982). "Studies of the mortality of A-bomb survivors. 7. Mortality, 1950–1978: Part 1. Cancer mortality," Radiat. Res. **90**, 395–432.

KENDALL, G.M., MUIRHEAD, C.R., MACGIBBON, B.H., O'HAGAN, J.A., CONQUEST, A.J., GOODILL, A.A., BUTLAND, B.K., FELL, T.P., JACKSON, D.A., WEBB, M.A., HAYLOCK, R.G.E., THOMAS, J.M and SILK, T.J. (1992). "Mortality and occupational exposure to radiation: First analysis of the National Registry for Radiation Workers," Br. Med. J. **304**, 220–225.

KRIEGEL, H., SCHMAL, W., GERBER, G. and STIEVE, E.F., Eds. (1986). *Radiation Risks to the Developing Nervous System* (Gustav Fischer Verlag, New York).

LAIRD, N.M. (1987). "Thyroid cancer from exposure to ionizing radiation: A case study in the comparative potency model," Risk Anal. **7**, 299–309.

LAND, C.E. and SINCLAIR, W.K. (1991). "The relative contributions of different organ sites to the total cancer mortality associated with low-dose radiation exposure," pages 31 to 57 in *Risks Associated with Ionising Radiations*, Annals of the ICRP **22** (Pergamon Press, Elmsford, New York).

MARTIN, R.H., HILDEBRAND, K., YAMAMOTO, J., RADEMAKER, A., BARNES, M., DOUGLAS, G., ARTHUR, K., RINGROSE, T. and BROWN, I.S. (1986). "An increased frequency of human sperm chromosomal abnormalities after radiotherapy," Mutat. Res. **174**, 219–225.

MARTIN, R.H., RADEMAKER, A., HILDEBRAND, K., BARNES, M., ARTHUR, K., RINGROSE, T., BROWN, J.S. and DOUGLAS, G. (1989). "A comparison of chromosomal aberrations induced by *in vivo* radiotherapy in human sperm and lymphocytes," Mutat. Res. **226**, 21–30.

MAYS, C.W. and SPIESS, H. (1983). "Epidemiological studies of German patients injected with ^{224}Ra," pages 159 to 166 in *Epidemiology Applied to Health Physics*, Proceedings of the Sixteenth Mid-Year Topical Meeting of the Health Physics Society, CONF. 830101 (National Technical Information Service, Springfield, Virginia).

MCKINNEY, P.A., ALEXANDER, F.E., CARTWRIGHT, R.A. and PARKER, L. (1991). "Parental occupations of children with leukaemia in West Cumbria, North Humberside, and Gateshead," Brit. Med. J. **302**, 681–687.

MEADOWS, A.T., GORDON, J., MASSARI, D.J., LITTMAN, P., FERGUSSON, J. and MOSS, K. (1981). "Declines in IQ scores and cognitive dysfunctions in children with acute lymphocytic leukaemia treated with cranial irradiation," Lancet ii, 1015–1018.

MILLER, R.W. and BLOT, W.J. (1972). "Small head size after in-utero exposure to atomic irradiation," Lancet ii, 784–787.

MILLER, R.W. and MULVIHILL, J.J. (1976). "Small head size after atomic irradiation," Teratology 14, 335–337.

MILLER, A.B., HOWE, G.R., SHERMAN, G.J., LINDSAY, J.P., YAFFE, M.J., DINNER, P.J., RISCH, H.A. and PRESTON, D.L. (1989). "Mortality from breast cancer after irradiation during fluoroscopic examinations in patients being treated for tuberculosis," New Engl. J. Med. 321, 1285–1289.

MOLE, R.H. (1974). "Antenatal irradiation and childhood cancer: Causation or coincidence?" Brit. J. Cancer 30, 199–208.

MONSON, R.R. and MACMAHON, B. (1984). "Prenatal x-ray exposure and cancer in children," pages 97 to 105 in Radiation Carcinogenesis: Epidemiology and Biological Significance, Boice, J.D., Jr. and Fraumeni, J.F., Jr., Eds. (Raven Press, New York).

MUIRHEAD, C.R. and DARBY, S.C. (1987). "Modeling the relative and absolute risks of radiation-induced cancers," J. R. Statist. Soc. A150, 83–118.

MÜLLER, H.J. (1927). "Artificial transmutation of the gene," Science 66, 84–87.

MULLER, J., KUSIAK, R.A. and RITCHIE, A.L. (1988). Modifying Factors in Lung Cancer Risk of Ontario Uranium Miners 1955–1981, Ontario Ministry of Labour Report (Ministry of Labour, Toronto).

NAS/NRC (1972). National Academy of Sciences/National Research Council, Committee on the Biological Effects of Ionizing Radiations. The Effects on Populations of Exposure to Low Levels of Ionizing Radiation, BEIR Report (National Academy Press, Washington).

NAS/NRC (1980). National Academy of Sciences/National Research Council, Committee on the Biological Effects of Ionizing Radiations. The Effects on Populations of Exposure to Low Levels of Ionizing Radiation: 1980, BEIR III (National Academy Press, Washington).

NAS/NRC (1988). National Academy of Sciences/National Research Council, Committee on the Biological Effects of Ionizing Radiations. Health Risks of Radon and Other Internally Deposited Alpha Emitters, BEIR IV (National Academy Press, Washington).

NAS/NRC (1990). National Academy of Sciences/National Research Council, Committee on the Biological Effects of Ionizing Radiations. Health Effects of Exposure to Low Levels of Ionizing Radiation, BEIR V (National Academy Press, Washington).

NCRP (1980). National Council on Radiation Protection and Measurements. Influence of Dose and its Distribution in Time on Dose-Response Relationships for Low-Let Radiations, NCRP Report No. 64 (National Council on Radiation Protection and Measurements, Bethesda, Maryland).

NCRP (1984). National Council on Radiation Protection and Measurements. *Evaluation of Occupational and Environmental Exposures to Radon and Radon Daughters in the United States*, NCRP Report No. 78 (National Council on Radiation Protection and Measurements, Bethesda, Maryland).

NCRP (1985). National Council on Radiation Protection and Measurements. *Induction of Thyroid Cancer by Ionizing Radiation*, NCRP Report No. 80 (National Council on Radiation Protection and Measurements, Bethesda, Maryland).

NCRP (1987). National Council on Radiation Protection and Measurements. *Recommendations on Limits for Exposure to Ionizing Radiation*, NCRP Report No. 91, (National Council on Radiation Protection and Measurements, Bethesda, Maryland).

NCRP (1991). National Council on Radiation Protection and Measurements. *Radon Exposure of the U.S. Population—Status of the Problem*, NCRP Commentary No. 6 (National Council on Radiation Protection and Measurements, Bethesda, Maryland).

NCRP (1993). National Council on Radiation Protection and Measurements. *Limitation of Exposure to Ionizing Radiation*, NCRP Report No. 116 (National Council on Radiation Protection and Measurements, Bethesda, Maryland).

NEEL, J.V., SCHULL, W.J., AWA, A.A., SATOH, C., KATO, H., OTAKE, M. and YOSHIMOTO, Y. (1990). "The children of parents exposed to atomic bombs: Estimates of the genetic doubling dose of radiation for humans," Am. J. Hum. Genet. **46**, 1053–1072.

NIH (1985). National Institutes of Health. *Report of the National Institutes of Health Ad Hoc Working Group to Develop Radioepidemiological Tables*, NIH Publication 85-2748 (U.S. Government Printing Office, Washington).

NIH (1989). National Institutes of Health. *Cancer Statistics Review, 1973–1986, Including a Report on the Status of Cancer Control 1989*, NIH Publication 87-2789 (U.S. Government Printing Office, Washington).

NRC (1985). Nuclear Regulatory Commission. *Health Effects Model for Nuclear Power Plant Accident Consequence Analysis. Part I: Introduction, Integration, and Summary. Part II: Scientific Basis for Health Effects Models*, NUREG/CR-4214 (U.S. Government Printing Office, Washington).

NRC (1989). Nuclear Regulatory Commission. *Health Effects Models for Nuclear Power Plant Accident Consequence Analysis. Low LET Radiation, Part II: Scientific Bases for Health Effects Models*, NUREG/CR-4214, Revision 1, Part II (U.S. Government Printing Office, Washington).

NRC (1991). Nuclear Regulatory Commission. *Health Effects Models for Nuclear Power Plant Accident Consequence Analysis. Modifications of Models Resulting from Recent Reports on Health Effects of Ionizing Radiation. Low LET Radiation, Part II: Scientific Bases for Health Effects Models*, NUREG/CR-4214, Revision 1, Part II, Addendum 1, LMF-132 (U.S. Government Printing Office, Washington).

NRPB (1988). National Radiological Protection Board. *Health Effects Models Developed from the 1988 UNSCEAR Report*, NRPB-R226 (National Radiological Protection Board, Oxon, England).

NRPB (1993). National Radiological Protection Board. *Estimates of Late Radiation Risks to the U.K. Population*, Documents of the NRPB, Volume 4, No. 4 (National Radiological Protection Board, Oxon, England).

OMRAN, A.R., SHORE, R.E., MARKOFF, R.A., FRIEDHOFF, A., ALBERT, R.E., BARR, H., DAHLSTROM, W.G. and PASTERNACK, B.S. (1978). "Follow-up study of patients treated by x-ray epilation for tinea capitis: Psychiatric and psychometric evaluation," Am. J. Public Health **68**, 561–567.

OTAKE, M. (1980). *Patterns in Cancer Mortality in the United States and Japan*, RERF Technical Report 13-79 (Radiation Effects Research Foundation, Hiroshima).

OTAKE, M., YOSHIMARU, H. and SCHULL, W.J. (1988a). *Severe Mental Retardation Among the Prenatally Exposed Survivors of the Atomic Bombing of Hiroshima and Nagasaki: A Comparison of the T65DR and DS86 Dosimetry Systems*, RERF Technical Report 16-87 (Radiation Effects Research Foundation, Hiroshima).

OTAKE, M., SCHULL, W.J., FUJIKOSHI, Y. and YOSHIMARU, H. (1988b). *Effect on School Performance of Prenatal Exposure to Ionizing Radiation in Hiroshima: A Comparison of the T65DR and DS86 Dosimetry Systems*, RERF Technical Report 2-88 (Radiation Effects Research Foundation, Hiroshima).

PIERCE, D.A. (1989). *An Overview of the Cancer Mortality Data on the Atomic Bomb Survivors*, RERF Commentary and Review 1-89 (Radiation Effects Research Foundation, Hiroshima).

PRESTON, D.L. (1993). Personal communication.

PRESTON, D.L. and PIERCE, D.A. (1988). "The effect of changes in dosimetry on cancer mortality risk estimates in the atomic bomb survivors," Radiat. Res. **114**, 437–466.

PRESTON, D.L., PIERCE, D. and VAETH, M. (1992–1993). "Neutrons and radiation risk: A commentary," page 5 in *RERF Update,* Volume 4, Issue 4 (Radiation Effects Research Foundation, Hiroshima).

PUSKIN, J.S. and YANG, Y. (1988). "A retrospective look at Rn-induced lung cancer mortality from the viewpoint of a relative risk model," Health Phys. **54**, 635–643.

RODVALL, Y., PERSHAGEN, G., HRUBEC, Z., AHLBOM, A., PEDERSEN, N.L. and BOICE, J.D., JR. (1990). "Prenatal x-ray exposure and childhood cancer in Swedish twins," Int. J. Cancer **46**, 362–365.

ROESCH, W.C., Ed. (1987). *U.S.–Japan Joint Reassessment of Atomic Bomb Radiation Dosimetry in Hiroshima and Nagasaki, Final Report* (Radiation Effects Research Foundation, Hiroshima).

ROMAN, E., BERAL, V., CARPENTER, L., WATSON, A., BARTON, C., RYDER, H. and ASTON, D.L. (1987). "Childhood leukaemia in the West Berkshire and Basingstoke and North Hampshire District Health Authorities in relation to nuclear establishments in the vicinity," Br. Med. J. **294**, 597–602.

RON, E. and MODAN, B. (1984). "Thyroid and other neoplasms following childhood scalp irradiation," pages 139 to 151 in *Radiation Carcinogenesis:*

Epidemiology and Biological Significance, Boice, J.D., Jr. and Fraumeni, J.F., Jr., Eds. (Raven Press, New York).

RON, E., MODAN, B., FLORO, S., HARKEDAR, I. and GUREWITZ, R. (1982). "Mental function following scalp irradiation during childhood," Am. J. Epidemiol. **116**, 149–160.

SCHATZKIN, A. and SLUD, E. (1989). "Competing risks bias arising from an omitted risk factor," Am. J. Epidemiol. **129**, 850–856.

SCHULL, W.J. and OTAKE, M. (1987). *Effect on Intelligence of Prenatal Exposure to Ionizing Radiation*, RERF Technical Report 7-86 (Radiation Effects Research Foundation, Hiroshima).

SCHULL, W.J., OTAKE, M. and NEEL, J.V. (1981). "Genetic effects of the atomic bombs: A reappraisal," Science **213**, 1220–1227.

SCHULL, W.J., OTAKE, M. and YOSHIMARU, H. (1988). *Effect on Intelligence Test Score of Prenatal Exposure to Ionizing Radiation in Hiroshima and Nagasaki: A Comparison of the T65DR and DS86 Dosimetry Systems*, RERF Technical Report 3-88 (Radiation Effects Research Foundation, Hiroshima).

SHIMIZU, Y., KATO, H., SCHULL, W.J., PRESTON, D.L., FUJITA, S. and PIERCE, D.A. (1987). *Life Span Study Report 11, Part 1. Comparison of Risk Coefficients for Site-Specific Cancer Mortality Based on the DS86 and T65DR Shielded Kerma and Organ Doses*, RERF Technical Report 12-87 (Radiation Effects Research Foundation, Hiroshima).

SHIMIZU, Y., KATO, H. and SCHULL, W.J. (1988). *Life Span Study Report 11, Part 2. Cancer Mortality in the Years 1950–85 Based on the Recently Revised Doses (DS86)*, RERF Technical Report 5-88 (Radiation Effects Research Foundation, Hiroshima).

SHIMIZU, Y., KATO, H., SCHULL, W.J., PRESTON, D.L., FUJITA, S. and PIERCE, D.A. (1989). "Studies of the mortality of A-bomb survivors. 9. Mortality, 1950–1985: Part 1. Comparison of risk coefficients for site-specific cancer mortality based on the DS86 and T65DR shielded kerma and organ doses," Radiat. Res. **118**, 502–524.

SHIMIZU, Y., KATO, H. and SCHULL, W.J. (1990). "Studies of the mortality of A-bomb survivors. 9. Mortality, 1950–1985: Part 2. Cancer mortality based on the recently revised doses (DS86)," Radiat. Res. **121**, 120–141.

SHORE, R.E. (1990). "Overview of radiation-induced skin cancer in humans," Int. J. Radiat. Biol. **57**, 809–827.

SHORE, R.E., ALBERT, R.E. and PASTERNACK, B.S. (1976). "Follow-up study of patients treated by x-ray epilation for tinea capitis; resurvey of post-treatment illness and mortality experience," Arch. Environ. Health **31**, 21–28.

SHORE, R.E., ALBERT, R.E., REED, M., HARLEY, N.H. and PASTERNACK, B.S. (1984). "Skin cancer incidence among children irradiated for ringworm of the scalp," Radiat. Res. **100**, 192–204.

SHORE, R.E., WOODARD, E., HILDRETH, N., DVORETSKY, P., HEMPELMANN, L. and PASTERNACK, B.S. (1985). "Thyroid tumors following thymus irradiation," J. Natl. Cancer Inst. **74**, 1177–1184.

SHORE, R.E., HILDRETH, N., WOODARD, E., DVORETSKY, P., HEMPELMANN, L. and PASTERNACK, B.S. (1986). "Breast cancer among women given x-ray therapy for acute postpartum mastitis," J. Natl. Cancer Inst. **77**, 689–696.

SINCLAIR, W.K. (1985). "Implications of risk information for the NCRP program," pages 223 to 237 in *Some Issues Important in Developing Basic Radiation Protection Recommendations*, NCRP Proceedings No. 6 (National Council on Radiation Protection and Measurements, Bethesda, Maryland).

SINCLAIR, W.K. (1992). "Radiation induced cancer risk, detriment and radiation protection," pages 275 to 281 in *Proceedings of International Conference on Radiation Effects and Protection Association* (Japan Atomic Energy Research Institute, Tokyo).

SINCLAIR, W.K. and PRESTON, D.L. (1987). "Revisions in the dosimetry of the A-bomb survivors at Hiroshima and Nagasaki and their consequences," pages 588 to 594 in *Radiation Research, Proceedings of the Eighth International Congress of Radiation Research* (Taylor and Francis, Philadelphia).

SMITH, P.G. and DOUGLAS, A.J. (1986). "Mortality of workers at the Sellafield plant of British Nuclear Fuels," Br. Med. J. **293**, 845–854.

STEWART, A. and KNEALE, G.W. (1970). "Radiation dose effects in relation to obstetric x-rays and childhood cancer," Lancet **i**, 1185–1188.

STORER, J.B. (1982). "Associations between tumor types in irradiated BALB/c female mice," Radiat. Res. **92**, 396–404.

STRAM, D.O. and MIZUNO, S. (1989). "Analysis of the DS86 atomic bomb radiation dosimetry methods using data on severe epilation," Radiat. Res. **117**, 93–113.

TOKUNAGA, M., LAND, C.E., YAMAMOTO, T., ASANO, M., TOKUOKA, S., EZAKI, H. and NISHIMORI, I. (1987). "Incidence of female breast cancer among atomic bomb survivors, Hiroshima and Nagasaki, 1950–1980," Radiat. Res. **112**, 243–272.

TSIATIS, A. (1975). "A nonidentifiability aspect of the problem of competing risks," Proc. Natl. Acad. Sci. USA **72**, 20–22.

ULLRICH, R.L., JERNIGAN, M.C., SATTERFIELD, L.C. and BOWLES, N.D. (1987). "Radiation carcinogenesis: Time-dose relationships," Radiat. Res. **111**, 179–184.

UNSCEAR (1972). United Nations Scientific Committee on the Effects of Atomic Radiation. *Ionizing Radiation: Levels and Effects*, 1972 Report to the General Assembly with annexes, Publication E.72.IX.17 (United Nations, New York).

UNSCEAR (1977). United Nations Scientific Committee on the Effects of Atomic Radiation. *Sources and Effects of Ionizing Radiation*, 1977 Report to the General Assembly, with annexes, Publication E.77.IX.1 (United Nations, New York).

UNSCEAR (1986). United Nations Scientific Committee on the Effects of Atomic Radiation. *Genetic and Somatic Effects of Ionizing Radiation*, 1986 Report to the General Assembly, with annexes, Publication E.86.IX.9 (United Nations, New York).

UNSCEAR (1988). United Nations Scientific Committee on the Effects of Atomic Radiation. "Annex F: Radiation carcinogenesis in man," pages

405 to 543 in *Sources, Effects and Risks of Ionizing Radiation*, 1988 Report to the General Assembly, with annexes, Publication E.88.IX.7 (United Nations, New York).

UPTON, A.C. (1991). "Risk estimates for carcinogenic effects of radiation," pages 1 to 29 in *Risks Associated with Ionizing Radiations*, Annals of the ICRP **22** (Pergamon Press, Elmsford, New York).

URQUHART, J.D., BLACK, R.J., MUIRHEAD, M.J., SHARP, L., MAXWELL, M., EDEN, O.B. and JONES, D.A. (1991). "Case-control study of leukaemia and non-Hodgkin's lymphoma in children in Caithness near the Dounreay nuclear installation," Brit. Med. J. **302**, 687–692.

VAN KAICK, G., WESCH, H., LÜHRS, H., LIEBERMANN, D., KAUL, A. and MUTH, H. (1989). "The German thorotrast study—report on 20 years follow-up," pages 98 to 104 in *Risks from Radium and Thorotrast*, BIR Report 21 (British Institute of Radiology, London).

WING, S., SHY, C.M., WOOD, J.L., WOLF, S., CRAGLE., D.L. and FROME, E.L. (1991). "Mortality among workers at Oak Ridge National Laboratory. Evidence of radiation effects in follow-up through 1984," JAMA **265**, 1397–1402.

YASHIN, A.I., MANTON, K.G. and STALLARD, E. (1986). "Dependent competing risks: A stochastic process model," J. Math. Biol. **24**, 119–140.

YOSHIMOTO, Y., KATO, H. and SCHULL, W.J. (1988). *Risk of Cancer Among In Utero Children Exposed to A-Bomb Radiation, 1950–84*, RERF Technical Report 4-88 (Radiation Effects Research Foundation, Hiroshima).

YOSHIMOTO, Y., NEEL, J.V., SCHULL, W.J., KATO, H., SODA, M., ETO, R. and MABUCHI, K. (1990). "Malignant tumors during the first two decades of life in the offspring of atomic bomb survivors," Am. J. Hum. Genet. **46**, 1041–1052.

The NCRP

The National Council on Radiation Protection and Measurements is a nonprofit corporation chartered by Congress in 1964 to:

1. Collect, analyze, develop and disseminate in the public interest information and recommendations about (a) protection against radiation and (b) radiation measurements, quantities and units, particularly those concerned with radiation protection.
2. Provide a means by which organizations concerned with the scientific and related aspects of radiation protection and of radiation quantities, units and measurements may cooperate for effective utilization of their combined resources, and to stimulate the work of such organizations.
3. Develop basic concepts about radiation quantities, units and measurements, about the application of these concepts, and about radiation protection.
4. Cooperate with the International Commission on Radiological Protection, the International Commission on Radiation Units and Measurements, and other national and international organizations, governmental and private, concerned with radiation quantities, units and measurements and with radiation protection.

The Council is the successor to the unincorporated association of scientists known as the National Committee on Radiation Protection and Measurements and was formed to carry on the work begun by the Committee.

The Council is made up of the members and the participants who serve on the scientific committees of the Council. The Council members who are selected solely on the basis of their scientific expertise are drawn from public and private universities, medical centers, national and private laboratories and industry. The scientific committees, composed of experts having detailed knowledge and competence in the particular area of the committee's interest, draft proposed recommendations. These are then submitted to the full membership of the Council for careful review and approval before being published.

127

The following comprise the current officers and membership of the Council:

Officers

President	CHARLES B. MEINHOLD
Vice President	S. JAMES ADELSTEIN
Secretary and Treasurer	W. ROGER NEY
Assistant Secretary	CARL D. HOBELMAN
Assistant Treasurer	JAMES F. BERG

Members

SEYMOUR ABRAHAMSON	THOMAS F. GESELL	WILLIAM A. MILLS
S. JAMES ADELSTEIN	ETHEL S. GILBERT	DADE W. MOELLER
PETER R. ALMOND	ROBERT A. GOEPP	GILBERT S. OMENN
LYNN R. ANSPAUGH	JOEL E. GRAY	LESTER J. PETERS
JOHN A. AUXIER	ARTHUR W. GUY	RONALD PETERSEN
JOHN W. BAUM	ERIC J. HALL	JOHN W. POSTON, SR.
HAROLD L. BECK	NAOMI H. HARLEY	ANDREW K. POZNANSKI
MICHAEL A. BENDER	WILLIAM R. HENDEE	GENEVIEVE S. ROESSLER
B. GORDON BLAYLOCK	DAVID G. HOEL	MARVIN ROSENSTEIN
BRUCE B. BOECKER	F. OWEN HOFFMAN	LAWRENCE N. ROTHENBERG
JOHN D. BOICE, JR.	DONALD G. JACOBS	MICHAEL T. RYAN
ANDRÉ BOUVILLE	A. EVERETTE JAMES, JR.	KEITH J. SCHIAGER
ROBERT L. BRENT	JOHN R. JOHNSON	ROBERT A. SCHLENKER
A. BERTRAND BRILL	BERND KAHN	ROY E. SHORE
ANTONE L. BROOKS	KENNETH R. KASE	DAVID H. SLINEY
PAUL L. CARSON	AMY KRONENBERG	PAUL SLOVIC
MELVIN W. CARTER	HAROLD L. KUNDEL	RICHARD A. TELL
JAMES E. CLEAVER	CHARLES E. LAND	WILLIAM L. TEMPLETON
FRED T. CROSS	JOHN B. LITTLE	THOMAS S. TENFORDE
GAIL DE PLANQUE	HARRY R. MAXON	RALPH H. THOMAS
SARAH DONALDSON	ROGER O. MCCLELLAN	JOHN E. TILL
CARL H. DURNEY	BARBARA J. MCNEIL	ROBERT L. ULLRICH
KEITH F. ECKERMAN	CHARLES B. MEINHOLD	DAVID WEBER
CHARLES M. EISENHAUER	FRED A. METTLER, JR.	F. WARD WHICKER
THOMAS S. ELY		MARVIN C. ZISKIN

Honorary Members

LAURISTON S. TAYLOR, *Honorary President*
WARREN K. SINCLAIR, *President Emeritus*

EDWARD L. ALPEN	R.J. MICHAEL FRY	JOHN H. RUST
WILLIAM J. BAIR	ROBERT O. GORSON	EUGENE L. SAENGER
VICTOR P. BOND	JOHN W. HEALY	LEONARD A. SAGAN
REYNOLD F. BROWN	PAUL C. HODGES	WILLIAM J. SCHULL
RANDALL S. CASWELL	GEORGE V. LEROY	J. NEWELL STANNARD
FREDERICK P. COWAN	WILFRID B. MANN	JOHN B. STORER
JAMES F. CROW	A. ALAN MOGHISSI	ROY C. THOMPSON
GERALD D. DODD	KARL Z. MORGAN	ARTHUR C. UPTON
PATRICIA W. DURBIN	ROBERT J. NELSEN	GEORGE L. VOELZ
MERRIL EISENBUD	WESLEY L. NYBORG	EDWARD W. WEBSTER
ROBLEY D. EVANS	CHESTER R. RICHMOND	GEORGE M. WILKENING
RICHARD F. FOSTER	HARALD H. ROSSI	HAROLD O. WYCKOFF
HYMER L. FRIEDELL	WILLIAM L. RUSSELL	

Currently, the following committees are actively engaged in formulating recommendations:

SC 1   Basic Radiation Protection Criteria
        SC 1-3 Collective Dose
        SC 1-4 Extrapolation of Risk from Non-Human Experimental
            Systems to Man
        SC 1-5 Uncertainty in Risk Estimates
SC 9   Structural Shielding Design and Evaluation for Medical Use of X
            Rays and Gamma Rays of Energies Up to 10 MeV
SC 16   X-Ray Protection in Dental Offices
SC 46   Operational Radiation Safety
        SC 46-8 Radiation Protection Design Guidelines for Particle
            Accelerator Facilities
        SC 46-9 ALARA at Nuclear Plants
        SC 46-10 Assessment of Occupational Doses from Internal Emitters
        SC 46-11 Radiation Protection During Special Medical Procedures
        SC 46-12 Determination of the Effective Dose Equivalent (and
            Effective Dose) to Workers for External Exposure to Low-LET
            Radiation
SC 57   Dosimetry and Metabolism of Radionuclides
        SC 57-2 Respiratory Tract Model
        SC 57-9 Lung Cancer Risk
        SC 57-10 Liver Cancer Risk
        SC 57-14 Placental Transfer
        SC 57-15 Uranium
        SC 57-16 Uncertainties in the Application of Metabolic Models
SC 63   Radiation Exposure Control in a Nuclear Emergency
        SC 63-1 Public Knowledge
SC 64   Radionuclides in the Environment
        SC 64-6 Screening Models
        SC 64-17 Uncertainty in Environmental Transport in the Absence
            of Site Specific Data
        SC 64-18 Plutonium
        SC 64-19 Historical Dose Evaluation
SC 66   Biological Effects and Exposure Criteria for Ultrasound
SC 69   Efficacy of Radiographic Procedures
SC 72   Radiation Protection in Mammography
SC 75   Guidance on Radiation Received in Space Activities
SC 77   Guidance on Occupational and Public Exposure Resulting from
            Diagnostic Nuclear Medicine Procedures
SC 84   Radionuclide Contamination
        SC 84-1 Contaminated Soil
        SC 84-2 Decontamination and Decommissioning of Facilities
SC 85   Risk of Lung Cancer from Radon
SC 86   Hot Particles in the Eye, Ear or Lung
SC 87   Radioactive and Mixed Waste
        SC 87-1 Waste Avoidance and Volume Reduction
        SC 87-2 Waste Classification Based on Risk
        SC 87-3 Performance Assessment
SC 88   Fluence as the Basis for a Radiation Protection System for
            Astronauts

SC 89   Nonionizing Electromagnetic Fields
         SC 89-1 Biological Effects of Magnetic Fields
         SC 89-2 Practical Guidance on the Evaluation of Human Exposure to Radiofrequency Radiation
         SC 89-3 Extremely Low-Frequency Electric and Magnetic Fields
SC 90   Precautions in the Management of Patients Who Have Received Therapeutic Amounts of Radionuclides
SC 91   Radiation Protection in Medicine
Ad Hoc Committee on the Embryo Fetus and Nursing Child
Ad Hoc Committee on Council Involvement in Public Decision Making

In recognition of its responsibility to facilitate and stimulate cooperation among organizations concerned with the scientific and related aspects of radiation protection and measurement, the Council has created a category of NCRP Collaborating Organizations. Organizations or groups of organizations that are national or international in scope and are concerned with scientific problems involving radiation quantities, units, measurements and effects, or radiation protection may be admitted to collaborating status by the Council. Collaborating Organizations provide a means by which the NCRP can gain input into its activities from a wider segment of society. At the same time, the relationships with the Collaborating Organizations facilitate wider dissemination of information about the Council's activities, interests and concerns. Collaborating Organizations have the opportunity to comment on draft reports (at the time that these are submitted to the members of the Council). This is intended to capitalize on the fact that Collaborating Organizations are in an excellent position to both contribute to the identification of what needs to be treated in NCRP reports and to identify problems that might result from proposed recommendations. The present Collaborating Organizations with which the NCRP maintains liaison are as follows:

American Academy of Dermatology
American Association of Physicists in Medicine
American College of Medical Physics
American College of Nuclear Physicians
American College of Occupational and Environmental Medicine
American College of Radiology
American Dental Association
American Industrial Hygiene Association
American Institute of Ultrasound in Medicine
American Insurance Services Group
American Medical Association
American Nuclear Society
American Podiatric Medical Association
American Public Health Association
American Radium Society

American Roentgen Ray Society
American Society of Radiologic Technologists
American Society for Therapeutic Radiology and Oncology
Association of University Radiologists
Bioelectromagnetics Society
College of American Pathologists
Conference of Radiation Control Program Directors
Electric Power Research Institute
Federal Communications Commission
Federal Emergency Management Agency
Genetics Society of America
Health Physics Society
Institute of Nuclear Power Operations
International Brotherhood of Electrical Workers
National Aeronautics and Space Administration
National Electrical Manufacturers Association
National Institute of Standards and Technology
Nuclear Management and Resources Council
Oil, Chemical and Atomic Workers Union
Radiation Research Society
Radiological Society of North America
Society of Nuclear Medicine
United States Air Force
United States Army
United States Coast Guard
United States Department of Energy
United States Department of Housing and Urban Development
United States Department of Labor
United States Environmental Protection Agency
United States Navy
United States Nuclear Regulatory Commission
United States Public Health Services
Utility Workers Union of America

The NCRP has found its relationships with these organizations to be extremely valuable to continued progress in its program.

Another aspect of the cooperative efforts of the NCRP relates to the Special Liaison relationships established with various governmental organizations that have an interest in radiation protection and measurements. This liaison relationship provides: (1) an opportunity for participating organizations to designate an individual to provide liaison between the organization and the NCRP; (2) that the individual designated will receive copies of draft NCRP reports (at the time that these are submitted to the members of the Council) with an invitation to comment, but not vote; and (3) that new NCRP efforts might be discussed with liaison individuals as appropriate, so that they might have an opportunity to make suggestions on new studies and related matters. The following organizations participate in the Special Liaison Program:

Australian Radiation Laboratory
Commissariat a l'Energie Atomique (France)
Commission of the European Communities
Defense Nuclear Agency
International Commission on Non-Ionizing Radiation Protection
Japan Radiation Council
National Radiological Protection Board (United Kingdom)
National Research Council (Canada)
Office of Science and Technology Policy
Office of Technology Assessment
Ultrasonics Institute (Australia)
United States Air Force
United States Department of Health and Human Services
United States Department of Transportation
United States Nuclear Regulatory Commission

The NCRP values highly the participation of these organizations in the Special Liaison Program.

The Council also benefits significantly from the relationships established pursuant to the Corporate Sponsor's Program. The program facilitates the interchange of information and ideas and corporate sponsors provide valuable fiscal support for the Council's program. This developing program currently includes the following Corporate Sponsors:

Amersham Corporation
Commonwealth Edison
Consumers Power Company
Duke Power Company
Eastman Kodak Company
EG&G Rocky Flats
Landauer, Inc.
3M
Public Service Electric and Gas Company
Southern California Edison Company
Westinghouse Electric Corporation

The Council's activities are made possible by the voluntary contribution of time and effort by its members and participants and the generous support of the following organizations:

Agfa Corporation
Alfred P. Sloan Foundation
Alliance of American Insurers
American Academy of Dermatology
American Academy of Oral and Maxillofacial Radiology
American Association of Physicists in Medicine

American Cancer Society
American College of Medical Physics
American College of Nuclear Physicians
American College of Occupational and Environmental Medicine
American College of Radiology
American College of Radiology Foundation
American Dental Association
American Healthcare Radiology Administrators
American Industrial Hygiene Association
American Insurance Services Group
American Medical Association
American Nuclear Society
American Osteopathic College of Radiology
American Podiatric Medical Association
American Public Health Association
American Radium Society
American Roentgen Ray Society
American Society of Radiologic Technologists
American Society for Therapeutic Radiology and Oncology
American Veterinary Medical Association
American Veterinary Radiology Society
Association of University Radiologists
Battelle Memorial Institute
Canberra Industries, Inc.
Chem Nuclear Systems
Center for Devices and Radiological Health
College of American Pathologists
Committee on Interagency Radiation Research and Policy Coordination
Commonwealth of Pennsylvania
Defense Nuclear Agency
Edison Electric Institute
Edward Mallinckrodt, Jr. Foundation
EG&G Idaho, Inc.
Electric Power Research Institute
Federal Emergency Management Agency
Florida Institute of Phosphate Research
Fuji Medical Systems, U.S.A., Inc.
Genetics Society of America
Health Effects Research Foundation (Japan)
Health Physics Society
Institute of Nuclear Power Operations
James Picker Foundation
Martin Marietta Corporation
National Aeronautics and Space Administration
National Association of Photographic Manufacturers
National Cancer Institute
National Electrical Manufacturers Association
National Institute of Standards and Technology
Nuclear Management and Resources Council
Picker International
Radiation Research Society
Radiological Society of North America

Richard Lounsbery Foundation
Sandia National Laboratory
Society of Nuclear Medicine
Society of Pediatric Radiology
United States Department of Energy
United States Department of Labor
United States Environmental Protection Agency
United States Navy
United States Nuclear Regulatory Commission
Victoreen, Inc.

Initial funds for publication of NCRP reports were provided by a grant from the James Picker Foundation.

The NCRP seeks to promulgate information and recommendations based on leading scientific judgment on matters of radiation protection and measurement and to foster cooperation among organizations concerned with these matters. These efforts are intended to serve the public interest and the Council welcomes comments and suggestions on its reports or activities from those interested in its work.

NCRP Publications

NCRP publications are distributed by the NCRP Publications Office. Information on prices and how to order may be obtained by directing an inquiry to:

NCRP Publications
7910 Woodmont Avenue
Suite 800
Bethesda, MD 20814-3095

The currently available publications are listed below.

NCRP Reports

No.	Title
8	*Control and Removal of Radioactive Contamination in Laboratories* (1951)
22	*Maximum Permissible Body Burdens and Maximum Permissible Concentrations of Radionuclides in Air and in Water for Occupational Exposure* (1959) [Includes Addendum 1 issued in August 1963]
23	*Measurement of Neutron Flux and Spectra for Physical and Biological Applications* (1960)
25	*Measurement of Absorbed Dose of Neutrons, and of Mixtures of Neutrons and Gamma Rays* (1961)
27	*Stopping Powers for Use with Cavity Chambers* (1961)
30	*Safe Handling of Radioactive Materials* (1964)
32	*Radiation Protection in Educational Institutions* (1966)
35	*Dental X-Ray Protection* (1970)
36	*Radiation Protection in Veterinary Medicine* (1970)
37	*Precautions in the Management of Patients Who Have Received Therapeutic Amounts of Radionuclides* (1970)
38	*Protection Against Neutron Radiation* (1971)
40	*Protection Against Radiation from Brachytherapy Sources* (1972)
41	*Specification of Gamma-Ray Brachytherapy Sources* (1974)
42	*Radiological Factors Affecting Decision-Making in a Nuclear Attack* (1974)
44	*Krypton-85 in the Atmosphere—Accumulation, Biological Significance, and Control Technology* (1975)

101 *Exposure of the U.S. Population from Occupational Radiation* (1989)
102 *Medical X-Ray, Electron Beam and Gamma-Ray Protection for Energies Up to 50 MeV (Equipment Design, Performance and Use)* (1989)
103 *Control of Radon in Houses* (1989)
104 *The Relative Biological Effectiveness of Radiations of Different Quality* (1990)
105 *Radiation Protection for Medical and Allied Health Personnel* (1989)
106 *Limit for Exposure to "Hot Particles" on the Skin* (1989)
107 *Implementation of the Principle of As Low As Reasonably Achievable (ALARA) for Medical and Dental Personnel* (1990)
108 *Conceptual Basis for Calculations of Absorbed-Dose Distributions* (1991)
109 *Effects of Ionizing Radiation on Aquatic Organisms* (1991)
110 *Some Aspects of Strontium Radiobiology* (1991)
111 *Developing Radiation Emergency Plans for Academic, Medical or Industrial Facilities* (1991)
112 *Calibration of Survey Instruments Used in Radiation Protection for the Assessment of Ionizing Radiation Fields and Radioactive Surface Contamination* (1991)
113 *Exposure Criteria for Medical Diagnostic Ultrasound: I. Criteria Based on Thermal Mechanisms* (1992)
114 *Maintaining Radiation Protection Records* (1992)
115 *Risk Estimates for Radiation Protection* (1993)
116 *Limitation of Exposure to Ionizing Radiation* (1993)
117 *Research Needs for Radiation Protection* (1993)
118 *Radiation Protection in the Mineral Extraction Industry* (1993)

Binders for NCRP reports are available. Two sizes make it possible to collect into small binders the "old series" of reports (NCRP Reports Nos. 8-30) and into large binders the more recent publications (NCRP Reports Nos. 32-118). Each binder will accommodate from five to seven reports. The binders carry the identification "NCRP Reports" and come with label holders which permit the user to attach labels showing the reports contained in each binder.

The following bound sets of NCRP reports are also available:

Volume I.   NCRP Reports Nos. 8, 22
Volume II. NCRP Reports Nos. 23, 25, 27, 30
Volume III. NCRP Reports Nos. 32, 35, 36, 37
Volume IV. NCRP Reports Nos. 38, 40, 41

(Titles of the individual reports contained in each volume are given above.)

NCRP Commentaries

No. | Title

1 *Krypton-85 in the Atmosphere—With Specific Reference to the Public Health Significance of the Proposed Controlled Release at Three Mile Island* (1980)

2 *Preliminary Evaluation of Criteria for the Disposal of Trans-uranic Contaminated Waste* (1982)

3 *Screening Techniques for Determining Compliance with Environmental Standards—Releases of Radionuclides to the Atmosphere* (1986), Revised (1989)

4 *Guidelines for the Release of Waste Water from Nuclear Facilities with Special Reference to the Public Health Significance of the Proposed Release of Treated Waste Waters at Three Mile Island* (1987)

5 *Review of the Publication, Living Without Landfills* (1989)

6 *Radon Exposure of the U.S. Population—Status of the Problem* (1991)

7 *Misadministration of Radioactive Material in Medicine—Scientific Background* (1991)

8 *Uncertainty in NCRP Screening Models Relating to Atmospheric Transport, Deposition and Uptake by Humans* (1993)

Proceedings of the Annual Meeting

No.   Title

1   *Perceptions of Risk*, Proceedings of the Fifteenth Annual
     Meeting held on March 14-15, 1979 (including Taylor
     Lecture No. 3) (1980)

3   *Critical Issues in Setting Radiation Dose Limits*, Proceed-
     ings of the Seventeenth Annual Meeting held on April
     8-9, 1981 (including Taylor Lecture No. 5) (1982)

4   *Radiation Protection and New Medical Diagnostic
     Approaches*, Proceedings of the Eighteenth Annual Meet-
     ing held on April 6-7, 1982 (including Taylor Lecture
     No. 6) (1983)

5   *Environmental Radioactivity*, Proceedings of the Nine-
     teenth Annual Meeting held on April 6-7, 1983 (including
     Taylor Lecture No. 7) (1983)

6   *Some Issues Important in Developing Basic Radiation Pro-
     tection Recommendations*, Proceedings of the Twentieth
     Annual Meeting held on April 4-5, 1984 (including Taylor
     Lecture No. 8) (1985)

7   *Radioactive Waste*, Proceedings of the Twenty-first Annual
     Meeting held on April 3-4, 1985 (including Taylor Lecture
     No. 9) (1986)

8   *Nonionizing Electromagnetic Radiations and Ultrasound*,
     Proceedings of the Twenty-second Annual Meeting held
     on April 2-3, 1986 (including Taylor Lecture No. 10)
     (1988)

9   *New Dosimetry at Hiroshima and Nagasaki and Its Implica-
     tions for Risk Estimates*, Proceedings of the Twenty-third
     Annual Meeting held on April 8-9, 1987 (including Taylor
     Lecture No. 11) (1988)

10   *Radon*, Proceedings of the Twenty-fourth Annual Meeting
     held on March 30-31, 1988 (including Taylor Lecture
     No. 12) (1989)

11   *Radiation Protection Today—The NCRP at Sixty Years*, Pro-
     ceedings of the Twenty-fifth Annual Meeting held on
     April 5-6, 1989 (including Taylor Lecture No. 13) (1990)

12   *Health and Ecological Implications of Radioactively Con-
     taminated Environments*, Proceedings of the Twenty-
     sixth Annual Meeting held on April 4-5, 1990 (including
     Taylor Lecture No. 14) (1991)

13   *Genes, Cancer and Radiation Protection*, Proceedings of the
     Twenty-seventh Annual Meeting held on April 3-4, 1991
     (including Taylor Lecture No. 15) (1992)

Lauriston S. Taylor Lectures

Symposium Proceedings

The Control of Exposure of the Public to Ionizing Radiation in the Event of Accident or Attack, Proceedings of a Symposium held April 27-29, 1981 (1982)

NCRP Statements

No.   Title

Other Documents

The following documents of the NCRP were published outside of the NCRP Report, Commentary and Statement series:

> *Somatic Radiation Dose for the General Population*, Report of the Ad Hoc Committee of the National Council on Radiation Protection and Measurements, 6 May 1959, Science, February 19, 1960, Vol. 131, No. 3399, pages 482–486
>
> *Dose Effect Modifying Factors In Radiation Protection*, Report of Subcommittee M-4 (Relative Biological Effectiveness) of the National Council on Radiation Protection and Measurements, Report BNL 50073 (T-471) (1967) Brookhaven National Laboratory (National Technical Information Service Springfield, Virginia)

The following documents are now superseded and/or out of print:

NCRP Reports

No.	Title
1	*X-Ray Protection* (1931) [Superseded by NCRP Report No. 3]
2	*Radium Protection* (1934) [Superseded by NCRP Report No. 4]
3	*X-Ray Protection* (1936) [Superseded by NCRP Report No. 6]
4	*Radium Protection (1938)* [Superseded by NCRP Report No. 13]
5	*Safe Handling of Radioactive Luminous Compound* (1941) [Out of Print]
6	*Medical X-Ray Protection Up to Two Million Volts* (1949) [Superseded by NCRP Report No. 18]
7	*Safe Handling of Radioactive Isotopes* (1949) [Superseded by NCRP Report No. 30]
9	*Recommendations for Waste Disposal of Phosphorus-32 and Iodine-131 for Medical Users* (1951) [Out of Print]
10	*Radiological Monitoring Methods and Instruments* (1952) [Superseded by NCRP Report No. 57]
11	*Maximum Permissible Amounts of Radioisotopes in the Human Body and Maximum Permissible Concentrations in Air and Water* (1953) [Superseded by NCRP Report No. 22]
12	*Recommendations for the Disposal of Carbon-14 Wastes* (1953) [Superseded by NCRP Report No. 81]

13 *Protection Against Radiations from Radium, Cobalt-60 and Cesium-137* (1954) [Superseded by NCRP Report No. 24]

14 *Protection Against Betatron-Synchrotron Radiations Up to 100 Million Electron Volts* (1954) [Superseded by NCRP Report No. 51]

15 *Safe Handling of Cadavers Containing Radioactive Isotopes* (1953) [Superseded by NCRP Report No. 21]

16 *Radioactive-Waste Disposal in the Ocean* (1954) [Out of Print]

17 *Permissible Dose from External Sources of Ionizing Radiation* (1954) including *Maximum Permissible Exposures to Man, Addendum to National Bureau of Standards Handbook 59* (1958) [Superseded by NCRP Report No. 39]

18 *X-Ray Protection* (1955) [Superseded by NCRP Report No. 26]

19 *Regulation of Radiation Exposure by Legislative Means* (1955) [Out of Print]

20 *Protection Against Neutron Radiation Up to 30 Million Electron Volts* (1957) [Superseded by NCRP Report No. 38]

21 *Safe Handling of Bodies Containing Radioactive Isotopes* (1958) [Superseded by NCRP Report No. 37]

24 *Protection Against Radiations from Sealed Gamma Sources* (1960) [Superseded by NCRP Reports No. 33, 34 and 40]

26 *Medical X-Ray Protection Up to Three Million Volts* (1961) [Superseded by NCRP Reports No. 33, 34, 35 and 36]

28 *A Manual of Radioactivity Procedures* (1961) [Superseded by NCRP Report No. 58]

29 *Exposure to Radiation in an Emergency* (1962) [Superseded by NCRP Report No. 42]

31 *Shielding for High-Energy Electron Accelerator Installations* (1964) [Superseded by NCRP Report No. 51]

33 *Medical X-Ray and Gamma-Ray Protection for Energies up to 10 MeV—Equipment Design and Use* (1968) [Superseded by NCRP Report No. 102]

34 *Medical X-Ray and Gamma-Ray Protection for Energies Up to 10 MeV—Structural Shielding Design and Evaluation Handbook* (1970) [Superseded by NCRP Report No. 49]

39 *Basic Radiation Protection Criteria* (1971) [Superseded by NCRP Report No. 91]

43 *Review of the Current State of Radiation Protection Philosophy* (1975) [Superseded by NCRP Report No. 91]

45 *Natural Background Radiation in the United States* (1975) [Superseded by NCRP Report No. 94]

NCRP Proceedings

No.   Title

Index